Journal of
COMMERCIAL
BIOTECHNOLOGY
http://CommercialBiotechnology.com

AF896116

Journal of
COMMERCIAL BIOTECHNOLOGY
http://CommercialBiotechnology.com

PUBLISHER

Journal of Commercial Biotechnology is published quarterly in Washington, DC by thinkBiotech LLC.

Journal of Commercial Biotechnology is available online at *http://CommercialBiotechnology.com*. Visit the journal's website for focus, scope, and policies; submission guidelines; sample papers; and staff contacts. The website may also be used to order subscriptions.

CORRESPONDENCE

Business correspondence and inquiries should be addressed to *editor@CommercialBiotechnology.com* or to thinkBiotech LLC, 3909 Witmer Rd Box 416, Niagara Falls, NY 14305.

CUSTOMER SERVICE AND SUBSCRIPTION INQUIRIES

Susbcription policies and rates are posted at *http://CommercialBiotechnology.com/about/subscriptions*.

Subscriptions may be purchased by following the above link, or by sending a check, money order, or credit card details to the correspondence address above. Purchase orders, invoice requests, and additional questions may be directed to *editor@CommercialBiotechnology.com*.

2012 Subscriptions		
Student	Digital	US$165
Individual	Digital	US$220
	Print + Digital	US$275
Small company < 100 employees	Digital	US$550
Institutional	Digital	US$1040
	Print + Digital	US$1100

ADVERTISING

A media kit with advertising rates is posted at *http://CommercialBiotechnology.com/JCB-mediakit.pdf*. Additional questions may be directed to *editor@CommercialBiotechnology.com*.

REPRINTS

For reprints of this journal please contact the publisher at the address above or at *editor@CommercialBiotechnology.com*.

PERMISSIONS

For queries relating to reproduction rights please contact the publisher at the address above or at *editor@CommercialBiotechnology.com*.

COPYRIGHT

Print ISSN: 1462-8732
Online ISSN: 1478-565X

Journal of
COMMERCIAL BIOTECHNOLOGY

VOLUME 18 NUMBER 2 APRIL 2012

Contents

Continued ...

Editorial

The Biotechnology Entrepreneurship Boot Camp: From lectern to printing press

Journal of Commercial Biotechnology (2012) 18, 3–4. doi: 10.5912/jcb.532

THIS ISSUE OF the Journal of Commercial Biotechnology focuses on the proceedings of the Seventh Annual Biotechnology Entrepreneurship Boot Camp held in conjunction with the Biotechnology Industry Organization (BIO) annual conference in Washington, DC in June, 2011.

The Biotechnology Entrepreneurship Boot Camp was launched for the 2005 BIO Annual Meeting in Philadelphia. The Boot Camp was originally designed as a program for CSOs but is now expanded in scope and design to address a broad range of issues for entrepreneurs more generally. The Boot Camp was created in response to the growing need in the managerial, scientific and academic community to learn about the necessary elements and skills to transform technology and invention into a viable company. The insight and energy required for entrepreneurial success can be developed by anyone motivated to do the following: think strategically, select projects and plan for expeditious and cost-effective management, understand the requirements of all the involved stakeholders, and oversee the essential components of the commercialization process.

The Boot Camp travels from year to year to each of the BIO Annual Meeting venues — a veritable "moveable feast." Previously, the Boot Camp was offered at BIO's annual meetings in Chicago in 2006 and 2010, Boston in 2007, San Diego in 2008, and Atlanta in 2009. The creation of the syllabus, the recruitment of faculty, and the faculty's extensive preparation suggested that wherever possible there should be core faculty, i.e., a portion of the faculty from the Philadelphia Boot Camp who would volunteer from year to year. This approach has the added benefit of improving the presentations and the material from year to year as the faculty themselves identify what works, as well as how to teach together. Each year, additional faculty members are recruited from the host region.

Correspondence: Stephen M. Sammut, Wharton School, University of Pennsylvania, US. E-mail: smsammut@wharton.upenn.edu

Over the seven years of the boot camp, over 500 entrepreneurs have attended and taken away a broad spectrum of insights from the faculty.

The Boot Camp was founded and co-chaired by Professors Arthur Boni of the Tepper School of Carnegie Mellon University, Stephen Sammut of the Wharton School and Burrill & Company, and Jeffrey Libson, Partner, Pepper Hamilton LLP and Lecturer at Wharton School Health Care Management Program. The law firm PepperHamilton has also served as the Boot Camp's sponsor since its inception.

In previous years, local Co-Chairs were:

Chicago, 2006

Panayiotis P. Constantinides, Ph.D, Principal, Biopharmaceutical & Drug Delivery Consulting

Elsie Quaite-Randall, Ph.D., MBA Executive Director of the Office of Research Contracts and IP, McMaster University, formerly of Office of Technology Transfer, Argonne National Laboratory

Boston, 2007

Robert Creeden, Partners HealthCare Systems, Inc., Managing Director, Center for Innovative Ventures

San Diego, 2008

Kurt A. May, MBA, Assistant Dean, Executive Development Center, Rady School of Management, University of California, San Diego

Duane J. Roth, Chief Executive Officer and Member of the Board of CONNECT.

Atlanta, 2009

Frank R. Hunt, MS MBA, President, SE bioStrategies, Inc., CEO, PNP Therapeutics, Inc.

Dennis P. Schafer, CEO, Life Science Management

Chicago, 2010

Patrick G. Morand, Managing Director, SWMF Life Science Fund

Charles B. Hoslet, Managing Director, UW-Madison Office of Corporate Relations

Allen J. Dines, Assistant Director, Office of Corporate Relations, University of Wisconsin

Washington, 2011

Elana Fine, Director of venture Investments, Dingman Center for Entrepreneurship, Robert H. Smith School of Business, University of Maryland

Toby Gordon, Sc.D., Associate Professor, The Johns Hopkins Carey Business School

Martha J. Connolly, Ph.D, Director, Maryland Industrial Partnerships, University of Maryland

This edition of JCB includes articles based on the sessions from the Seventh Annual Boot Camp. The sequence of articles opens with the Keynote Address delivered by Dr. David C. U'Prichard, President and CEO, Druid Consulting, LLC, General Partner, Druid BioVentures and Former Chairman, Research and Development, GlaxoSmithKline on *New paradigms in drug R&D: A personal perspective.*

Arthur Boni, Ph.D., next addresses the issues of technology assessment in his article *Project, product or company.* This paper lays out a framework for determining not only the integrity of the technology but a determination of whether it has enough critical mass around which to form a company.

It is never too early to assess how the market will respond to products, especially as it relates to pricing and reimbursement. In a paper on *The basics of coverage, coding, and reimbursement for new medical devices and diagnostics* the fundamentals of pharmaceutical pricing and reimbursement strategies are explored by Robert Wanerman, JD, a partner at the law firm of Epstein, Becker & Green, PC.

James G. Kenimer, Ph.D., President & CEO, Biologics Consulting Group, Inc. and Jim Ackland, President, Global Biosolutions have provided an article on *Transition from the lab to the clinic — Regulatory considerations.* This article provides specific insight into planning for FDA regulations in light of strategy, financial needs, and the concerns of prospective partners and investors.

Arthur Boni follows this paper with two separate treatments. One on the ins and outs of building, developing and maintaining the management team and the other on the best approaches to writing the business plan and the pitch book.

Biotechnology companies often need guidance in working with the media. Moira Gunn, Ph.D., well known as the host of BioTech Nation on National Public Radio has provided her insights on the best approaches for relating to the media.

The next sheaf of articles is on the subject of partnering. Several speakers have converted their lectures into articles. They are: Benjamin Chen, PhD., Managing Partner, Ignatius Transaction Partners on the preparation of the partnering case and the best use of intermediaries. James Foley, Ph.D., CEO, Aqua Partners and former head of Business Development, BristolMyers Squibb provides the corporate perspective on the licensing process. His article is followed by Steve Fergusson, MBA, CLP, Deputy Director, Licensing and Entrepreneurship, National Institutes of Health addressing the finer points of working with government technology and developing licensing relationships with the US National Institutes of Health. The series on partnering is concluded by an article on working with universities by Wesley D. Blakeslee, JD, Executive Director, Technology Transfer, Johns Hopkins University.

The Boot Camp closes with a session on capitalization of the venture. Stephen Sammut offers two papers — the first on the venture capital due diligence process and a second on valuation methods used by venture capitalists. Ashley J. Stevens, D.Phil (Oxon), CLP, Special Asssitant to the Vice President for Research Technology Development and Senior Research Associate, ITEC concludes with a thorough piece on capitalization tables.

The Boot Camp historically includes a session on intellectual property strategy conducted by Raymond A. Miller, JD, a Partner, at Pepper Hamilton LLP and Kathryn Doyle, Ph.D., JD Partner and Chair, Riverside Law LLP. In between the 2011 Boot Camp and the publication of this issue, the United States Government passed the Leahy-Smith America Invents Act which has profound implications for the management of biotechnology patents. As of press time, regulations and Patent Office procedures have not been promulgated, so Ray and Kathryn asked that publication of their article be delayed so that it can be revised to address the new issues. Look for it in a future edition of JCB.

As the Co-Chairs of the Boot Camp, we invite any comments and suggestions from the readers on these proceedings and look forward to meeting you at the next Boot Camp, scheduled for June 18 and 19 2012 at the BIO Conference in Boston.

Stephen M. Sammut
Senior Fellow, Health Care Management
Wharton School,
University of Pennsylvania

Arthur A. Boni, Ph.D.
John R. Thorne Chair of Entrepreneurship, Tepper School of Business;
Director, Donald H. Jones Center for Entrepreneurship, Carnegie Mellon University

Keynote

New paradigms in drug R&D: A personal perspective

Received: November 8 2011

David C U'Prichard

is currently venture investor and director of 6 biotechnology companies. Formerly CEO, 3-Dimensional Pharmaceuticals, Chairman R&D Smithkline Beecham, EVP and Global Director of Research, Zeneca.

ABSTRACT

The author discusses the recent productivity problems in the pharmaceutical industry in the context of his 30 year career, and the current responses of the industry driven strongly by disaggregation of the historic R&D model, new fluidites of capital access, and the impacts of genomics and globalization.

Journal of Commercial Biotechnology (2012) 18, 5–12. doi: 10.5912/jcb.502

Keywords: pharma R&D; venture capital; new R&D models; globalization; virtualization

INTRODUCTION

I HAVE LUCKILY BEEN able to participate in the enterprise of new drug research and development from several different perspectives — academic research scientist, senior pharmaceutical R&D executive, biotechnology company CEO and venture capitalist. I trained in Britain and the U.S. as a pharmacologist, attracted to the discipline as a young scientist specifically because it applied intellectually compelling pure science, chemistry and biology, to directly benefit people's health and wellbeing. My very fortunate career has included opportunities to travel the world, meeting scientists and business people with similar interests from many different cultures. The worldwide enterprise of the discovery and bringing to the market of new drugs is in a tremendously exciting, fluctuating phase; old business models that have sustained a highly profitable industry in the U.S. and Europe are breaking down even as our knowledge of disease pathology and possible sites of new drug intervention has exploded. The enterprise has become vastly more globally interconnected as excellent scientists and doctors in Asia and South America contribute their particular medical traditions and experience. Whatever the woes of the established pharma industry in the West, the need and desire to develop new, better medicines remains very strong around the world, supported by new and unexpected sources of financial capital. I offer here a personal examination of the current state of this great enterprise, the evolving response of the drug makers and their financial backers to the impact of the unprecedented advances in biology, chemistry and informatics, and of globalization in the current, constrained economic environment.

Correspondence: David C U'Prichard, Druid Consulting LLC, US. E-mail: dcuprichard@druidconsult.com

THE GLOBAL PHARMACEUTICAL INDUSTRY'S PRODUCTIVITY PROBLEM

After a career as an academic researcher, and helping build Nova Pharmaceuticals in Baltimore, MD in the early 1980s, I joined the U.S. pharmaceuticals division of ICI, the UK-based chemical conglomerate — today's AstraZeneca. Over the previous quarter century, the pharma industry had become immensely profitable through the development of small molecules derived from industrial chemistry, to treat bacterial infections and the symptoms of many illnesses, protected by patents and the exclusive right of sale. Net margins were typically 30% or better. The less price-controlled U.S. market essentially subsidized the risky, expensive global enterprise of R&D, on which companies routinely spent 15-20% of their gross revenue. The prevailing competitive philosophy was to hire the brightest laboratory chemists and biologists from the top academic institutions and let them loose in the back lab to discover whatever they could, "throwing their compounds over the fence" for development and commercialization colleagues to exploit as best they could.

The alliance of brains and serendipity was the order of the day. At ICI in particular, the influence of Sir James Black (discoverer of the beta-blockers for cardiovascular disease at ICI, later H2-antagonists for ulcer therapy at Smith Kline & French, and ultimately Nobel laureate) reigned. Black's view was that "small science" guided by intellectual curiosity and uncontaminated by commercial considerations — small teams of closely knit chemists and pharmacologists — provided the scientific culture that ultimately afforded best returns on investment. The most important indices of activity of a potential new drug in the laboratory, and the most useful models for screening chemical analogs to optimize activity were response changes in function, in isolated organs and tissues, and animals *in vivo*; he was concerned about the artificiality of simpler *in vitro* biochemical screens, and he was unsympathetic to the theory that the best candidate molecules for further development could be found by science based on the laws of large numbers and statistical probabilities — large scale, high throughput, screening platforms coupled to combinatorial synthesis techniques that could synthesize and screen hundreds of thousands of simple analogs, that first became available in the late 1980s. In those days we were also attracted to "rational drug design" using classic enzyme and receptor theory, and pharmacophore models derived from structure-activity relationship data.

There was only a relatively rudimentary attempt to apply business portfolio ROI and business process models with any rigor to the more linear business of drug development, and certainly not to the looser, less rule-bound, more cyclical, discovery business. In the early 1990s, managing a group of 1000 discovery scientists at ICI Pharmaceuticals, I felt compelled to apply this thinking, to be competitive against larger groups, and was one of the first managers to institute more robust project and portfolio management processes. This was heretical at the time, the belief being that many excellent scientists would be driven away because of a presumed loss of "scientific freedom." The paradox became apparent over subsequent years that scientists recognize the benefit of operating freely within a robust, business-driven investment framework.

The era of ultra high throughput screen and very large scale combinatorial chemistry in the 1990s did not however increase pharmaceutical company R&D productivity as measured by new drug compounds entering the market with superior therapeutic profiles. In addition to this "big science" approach to new small molecule drugs, there was the advent of "biologics," first generation recombinant human proteins and monoclonal antibodies, which became a solid and expanding feature of the therapeutic marketplace after a slow start. Biologics initially were the raison d'être and sole province of the small pioneer biotechnology companies. Injectable biologics, limited to the hospital setting, were perceived to have medical and investment advantages: superior safety, and a faster track through clinical development, driven by similarity to natural body constituents. A theory also developed that small biotech drug discovery was somehow "smarter" and more fertile due to less bureaucratic overlay and quicker decision making, and therefore inherently more productive. As the enterprise of biologics R&D has matured, and big pharma has thoroughly assimilated biologics R&D alongside its small molecule heartland through internal growth or acquisition of products and companies, it is clear that some of these benefits do indeed accrue to monoclonal antibody product development, but by and large this theory has not been validated; there is no fundamental difference in R&D productivity between big pharma and small biotech.

Through the 1990s and 2000s, the global pharma companies increased their investment in R&D at a compound annual growth rate of 13%, much greater than inflation, but the number of new drugs getting to market each year in the U.S. remained static at about 20-25 per year. Inverting the relationship, the cost of developing an NCE (new chemical entity), including the cost of failures, went from $200M in 1980 to at least $1400M in 2010. Thus for many years now, "big pharma" has been perceived by the financial markets and industry analysts to be suffering declining productivity, with a "broken" R&D model (Figure 1). Many reasons can be adduced:

1. For most diseases, relief of overt and signs and symptoms with drug therapies represents the low hanging fruit, that the industry picked over for 30+ years with increasing success, such that achieving meaningful incremental benefit now over the standard of care has become harder and harder, across many diseases ranging from hypertension, heart failure, asthma, depression, schizophrenia, type 2 diabetes and bacterial infection.
2. The much more difficult goal that the drug R&D industry has set itself in the last twenty years is to delay or stop the underlying progression of degenerative disease. The symptoms of most illnesses outside infection are linked to underlying pathologies that are exacerbated with time. To modify the progression of disease with new drugs requires however a far more detailed understanding of the underlying genetics and pathobiology. The explosion of biological knowledge makes this much more attainable today compared to 10 years ago. However, to achieve this

understanding exhaustively enough, across many different disease targets and drug projects, to avoid mistakes and minimize risk due to incomplete knowledge of the pathobiology, requires a financial investment beyond the capacity of even the largest drug companies, and the patience of their investors. Hence, even with large year-over-year increases in R&D financial support, the odds of technical and commercial success have decreased in the last 20 years.

3. Regulatory agencies in the U.S. and Europe, animated by safety and general benefit/risk concerns, have required more clinical trial data for new drug applications, with larger numbers of patients needed in trials to demonstrate benefit over increasingly acceptable standards of care, and treatment in trials for a longer period to measure appropriate clinical endpoints that indicate whether the drug modifies the disease course. These factors collude to continually increase the cost of clinical development, usually at the expense of discovery budgets in an era of market-imposed R&D spending limits, which in turn militates against the effort to de-risk a compound before it enters clinical trials.
4. Although drugs remain a minor part of the bloated U.S. healthcare budget, drugs costs are much more visible, and controllable. There is an irresistible secular trend in the U.S. towards greater control of drug prices, towards levels achieved by government regulation in Europe, and this trend wil be further exacerbated by globalization. Again, this will negatively impact the R&D budget.

In sum, drug R&D (both pharma and biotech) has become less productive due to (a) greater expense to successfully execute a project with assured commercial value, plus (b) expansion of R&D budgets is not commensurate with the costs required to competitively extract benefit from the biology knowledge explosion. High single digit sector growth expectations promulgated to the markets ten years ago are now falling far short, as many billion-dollar-plus "blockbusters" are going off patent, and there is a "revenue gap" today of more than $100BN. These pressures have caused stock markets to

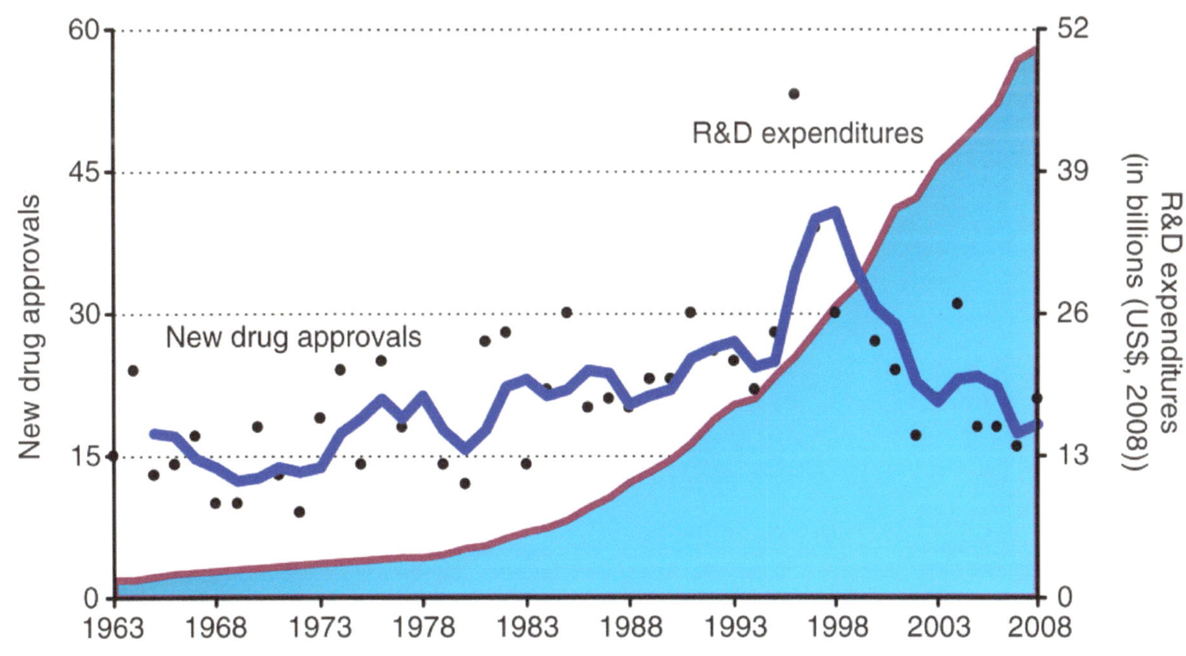

Figure 1: New drug approvals vs. pharmaceutical R&D expenditures

New drug approvals (dots), and pharmaceutical R&D expenditures (shaded area), in the United States from 1963 to 2008. R&D expenditures are presented in terms of constant 2008 dollar value. The trend line is a 3-year moving average. The source of drug approval data is the Tufts Center for the Study of Drug Development (CSDD). The source of R&D expenditure data is the Pharmaceutical Research and Manufacturers of America; Industry Profile 2009; conversion of actual expenses to constant dollars was performed by Tufts CSDD. *Only 3 in 10 new products generate revenues equal to or greater than average industry R&D costs.*

devalue the pharma sector over the last 5 years, leading to a cost-constraint response. However, as my then boss Tom McKillop pointed out in the early 1990s, when the industry was in the throes of "Hillarycare," the pharmaceutical industry is immensely profitable compared to other sectors, with at that time a 30% net margin. McKillop's point was that all those trends will decrease net margin, but pharma will nonetheless remain very profitable, because populations are increasing and aging, medical need generally is ever on the rise, around the world medicines produced by pharma will become more affordable to rising middle class populations, and around the world capital will find it attractive to invest in the drug R&D enterprise.

THE TECHNICAL EVOLUTION OF DRUG DISCOVERY AND DEVELOPMENT IN THE LAST 15 YEARS

From 1960 to 2000, the pharmaceutical industry accumulated about 500 biochemical protein "targets" (enzymes, receptors, transport proteins) that showed promise as loci of drug intervention. These targets became "validated" as clinical trials and subsequent market experience showed that drugs working at these sites achieved more or less the predicted benefit. The sequencing of the human genome ten years ago revealed the existence ultimately of 20-25,000 genes, and many more variant protein gene products, and in recent years it has become apparent that the non-coding portion of the genome is active and contains many drug targets. Thus the possible universe of relevant interventions is enormously larger than 500, and one estimate today is 5000 "druggable" targets. Our understanding of complex intracellular pathways, that cascade signaling at the cell surface to changes in gene expression and protein production, has also grown enormously in the last decade, but the great complexity and interplay of these systems is militates against the facile determination of the therapeutically optimal target (Figure 2).

The current task of R&D is to trace the best path from vast arrays of "omics" data, through cellular pathways, to cell processes such as apoptosis, angiogenesis, immune control and inflammatory degenerative responses, to ultimately impinge upon progressive disease responses in, for example, cancer, asthma and atherogenesis. The availability of large scale, ever-cheaper, sequencing and "omics" platforms is a two-edged sword. On the chemistry side, we can capture the vast expanse of chemistry space more readily with an enormous number of small "druggable" molecular structures that are amenable to combinatorial synthesis, that are available to "fit" target macromolecules. In addition, "rational design" has been enhanced through affordable techniques to co-crystallize macromolecular targets with many small molecules, and improved software allows for orders of magnitude more "virtual docking experiments" to be performed.

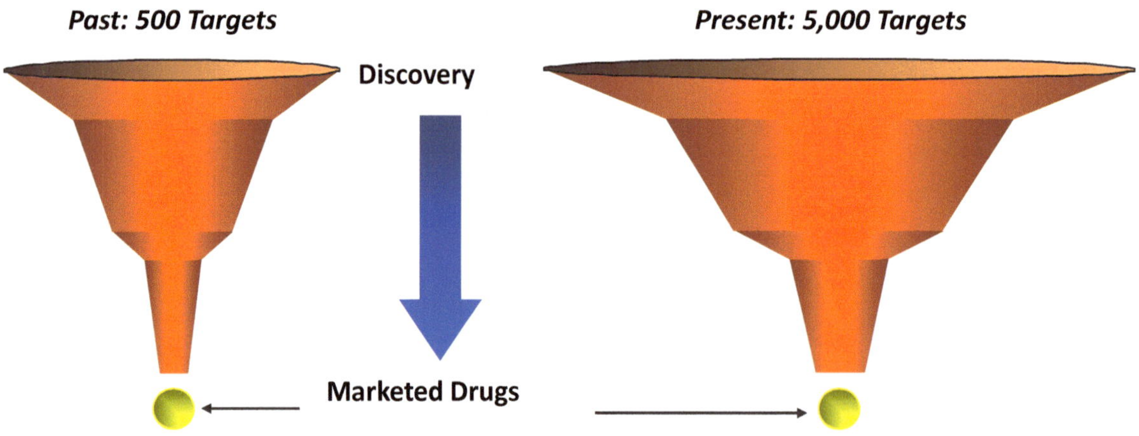

Figure 2: The pharmaceutical R&D funnel — No more low-hanging fruit
Huge influx of new targets from genomics
"Genomics Targets" incompletely validated
Lead Generation and Target Validation bottleneck – cost-effectiveness problem
Can cost $30-50 million to obtain POC
Personalized medicine is likely to fragment the market
High attrition accounts for high R&D costs: Failure consumes 75% of costs, and most of the "cost of failure" occurs in the laboratory, before the drug gets into the clinic

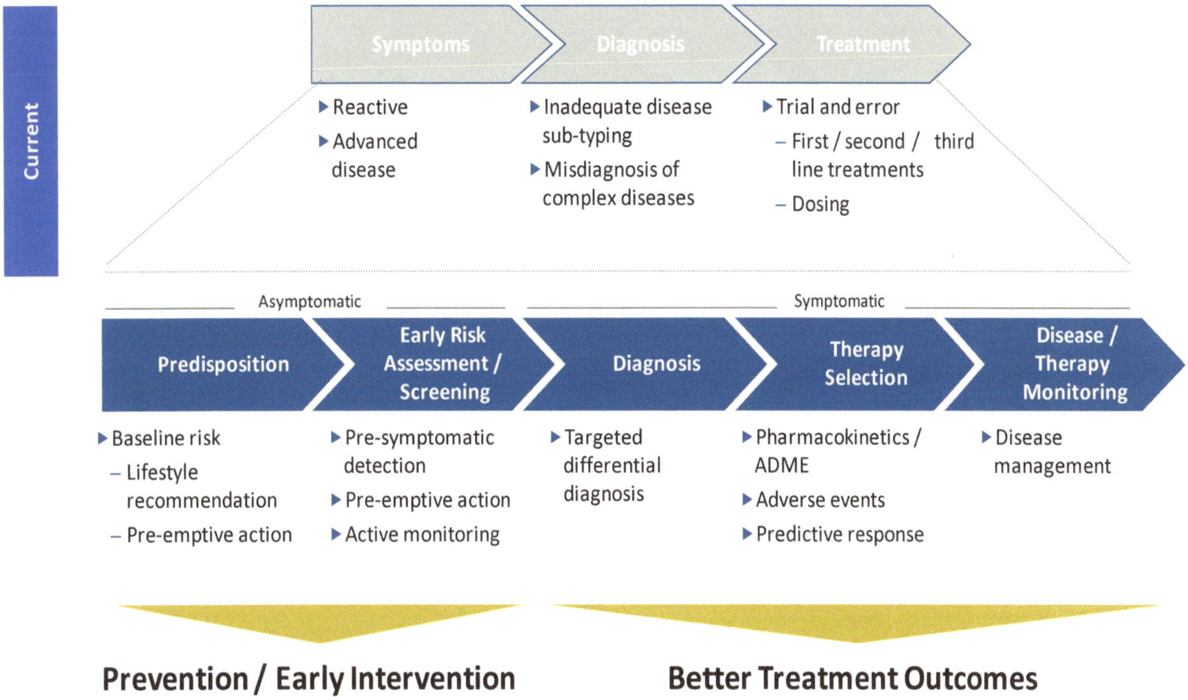

Figure 3: Molecular medicine changing the way medicine is (and will be) practiced
Courtesy Life Technologies, Inc.

Thus both "high throughput serendipidity" and "rational design" have been facilitated greatly in the last decade. However, now we have a data avalanche, and a logjam of interpretation, constraining in terms of time and money the power of the tools for drug discovery. The rush of technology has led to the crush of data. It is not surprising that there has been a recent sentiment in the industry to return to the James Black days of heuristic "small science"!

Cheap sequencing of individual genomes, discovery of individual variation in disease propensity, typing different molecular characteristics of a disease that superficially carries the same phenotype in many people, and understanding individual variation in efficacy and side-effects due to the body's different handling of drugs, is driving us steadily towards the era of "personalized medicine." The development of a "companion molecular diagnostic" alongside the new drug, to clarify the patient population for whom the drug is most beneficial, is fast becoming a *sine qua non* for some diseases such as cancer. There is a general requirement for the development of biomarkers that can impart confidence about drug-target engagement in patients, and enable a read on the presumptive efficacy of the drug, especially in a disease where clinical end-point data may mature very slowly. This new era is rapidly transforming yet further the drug R&D process. I sit on the board of Life Technologies, Inc., a company in the forefront of medical genomics, and am thus able to witness the impact of new, cheap, comprehensive, quantitative technologies at first hand, especially in DNA sequencing (Figure 3).

THE EVOLUTION OF R&D MANAGEMENT

Drug R&D is a process of sequential investment in a series of experiments, in the laboratory and then in the clinic, to further confirm efficacy and safety of the new drug and thereby reduce financial risk — the process of "validation." As accumulated data steadily reduces the risk of failure, the "present value" of the project increases. The difficult task of a research leader is to enable his organization to transform in-coming basic scientific data pertaining to targets, cell biology etc, into out-going packages of real drug information that fit the company's medical, commercial and financial frameworks. No small challenge, especially when dealing with a large R&D portfolio, many projects at different stages of evolution, aimed at a plethora of clinical targets, that furthermore need to be anchored by the best forward view of the commercial value of each project, taking account of all of the present and future competitive activity! At Zeneca in the late 1990's, we developed an ROI paradigm for estimating the return from long-term investment in a disease area like asthma. We started from a number of different clinical profiles ("Therapeutic Target Profiles," TPPs) that were worthwhile and had estimable future value. We then determined, with the best available knowledge, the key biological profiles translating to these TPPs with minimum risk. Working back, we agreed the molecular targets that, if amenable to successful engagement, would translate with minimum risk to the correct biology. The cost over years to extract this value for the disease area could be more or less calculated, and we had an ROI model for asthma as a company strategic investment, that we could then compare to other disease areas. Commercially and financially driven organization of discovery activities continues, I believe, to be cost-effective and doesn't inhibit creativity, but is very difficult to sustain in a large organization, which is why many large pharmas have consciously disaggregated their unitary R&D organizations to smaller, more independent, therapeutic units, that in principle compete with each other for resource.

In addition, the research manager must consider what continuing investment his organization can afford, to sustain the core underpinning technical platforms accessed by each project team. I believe a critical level of continual investment in new technology platforms is essential to maximize individual projects' ROI. Many small biotech companies are ultimately uncompetitive because they do not have such "table stakes" capital resource. However the advent of new technologies is never ending, and even the largest organizations need to find a "happy investment medium." Furthermore, we don't know where we are on the curve of technology-driven productivity improvement

PHARMA ADAPTING — VIRTUALIZATION AND GLOBALIZATION

Faced with enormous challenge today, from the explosion of biological and medical information, and from cost constraints imposed by the financial markets, the major pharmaceutical companies in the U.S. and Europe are seeking to spread the cost and risk of R&D, at the price of reward-sharing, in many different ways.

1. Faced with very large market opportunities in the primary care market, pharmas are increasingly inclined to partner with each other to burden share.
2. The historic relationship is evolving between pharmas and contract suppliers such as clinical research organizations. As CROs become themselves become "brains" as well as "hands," the relationship is evolving to a partnership of equals, with alliances being formed.
3. Pharma out-sourcing has now willingly moved upstream to the discovery component, to smaller players such as biotech companies, venture-backed groups of managers and even academic centers; the latter have steadily improved their own knowledge and competency to conduct the earliest part of drug discovery.. Typically, the pharma company will take an option on early external projects, reeling them in at IND or subsequent clinical proof-of-concept stage. Some senior pharma executives still maintain that the "quality" of research done in small external units is superior to their own shop, but fundamentally the rationale for discovery outsourcing in this era is that competitive discovery is now higher risk, and risk and cost should be defrayed. The rewards for success are great, therefore such profit-sharing is OK.
4. As biomarker development and companion diagnostics assume ever greater importance, pharma companies seem reluctant to internalize these new business requirements, but seek instead to partner with specialist companies in these areas.

Thus the large pharma companies may be evolving out of monolithic ownership of the entire R&D/commercial value chain, instead focusing on their chief competencies to ensure drug approval and maximize commercialization, partnering with other players to share the burden of reducing technical risk. My friend and predecessor as chairman of R&D at Smithkline Beecham, George Poste, has envisioned the "virtualization of the pharmaceutical industry," i.e. the dissolution of the global pharmaceutical monolith, and a re-integration of a skein of companies, joint ventures, academic centers and contractors, underpinned by rapidly growing open-source datasets — a vision I readily concur with (Figure 4).

In this new environment, the R&D of new, better drugs, hitherto the monopoly of the Western pharmaceutical companies, will "go global." Scientists and drug developers in Asia and South America are as inherently creative and innovative as their counterparts in the U.S. and Europe. We have seen rapid growth of local academic institutions and research centers, accompanied by the increasing return of Western-trained and educated scientists and managers; there is now no shortage of knowledge and talent in geographies with still comparatively low cost bases. Until costs globally equilibrate, there will be a strong impetus to conduct R&D, to the highest scientific and regulatory standards, increasingly in Asia and South America. I have been fortunate in recent years to be an advisor and director of two Indian companies, Advinus Therapeutics and Ocimum Biosolutions, whose

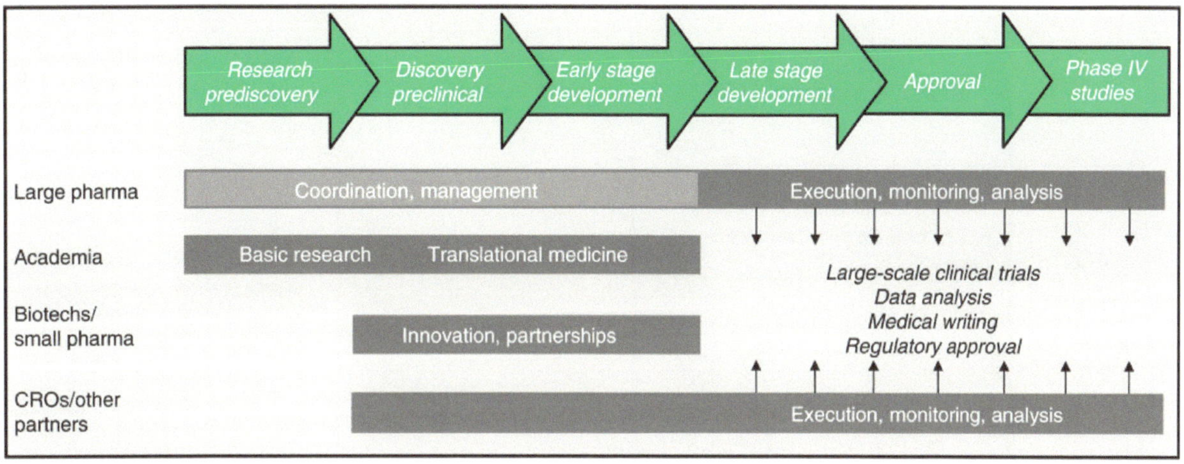

Figure 4: FIPNet (fully integrated pharmaceutical network) model of drug development
Shift from a fully integrated pharmaceutical company model of R&D, in which a sponsor "owns" the entire drug development process from synthesis to marketing, toward a networked model of innovation, sometimes referred to as a fully integrated pharmaceutical network, or FIPNet. FIPNets engage all the major stakeholders in the drug development process, melding the core competencies of each component to leverage capabilities, enhance efficiency, and boost output
FROM: Deconstructing the Drug Development Process: The New Face of Innovation, 2010, KI Kaitin, Tufts Center for the Study of Drug Development, Tufts University, Boston, Massachusetts, USA.

activities in internal R&D and contract services are second to none.

The ultimate vision is that of a global transformation of the entire industry. Digitization driven, the dissemination of cheap information will grow around the world, leading to weakened intellectual property structures, and the widespread availability of cheap generic drugs. Markets will segment. Conversely, globalization of the industry will be driven by the growth of Asian and South American health care consumers, rich and poor, and the concomitant proliferation of a variety of local health care services. The confluence of more education, more deregulation, rising demand and rising supply will induce long term growth of the market for drugs, accompanied by downsizing and specialization of corporations. The future is different, but bright.

VENTURE INVESTING TODAY IN DRUG R&D — CONVERGENCE WITH PHARMA'S NEEDS

The post-2008 constrained financial environment has been inimical to the classical biotech venture fund model. The investors in biotech funds (limited partners) are less risk tolerant, and LP oversight of fund investments has increased. Weaker funds have disappeared, and some of the strongest VCs have had to lower their funding targets as they close on new funds. As funds have become more capital constrained, their ability, even in syndication, to finance completion of many Phase II clinical trial programs, i.e. to enable their portfolio companies' assets to be "Phase III -ready," a status desired by more pharmas, is often shaky. In essence, pharmas are requiring that venture capital takes more risk out, especially of large primary care products that will still require hundreds of millions of dollars spent in further development to product launch. For portfolio biotech companies, this lack of alignment can spell stasis and death. I term this the major "back end problem" for venture backed biotech (Figure 5).

A counterpart "front end problem" for the VCs is that the traditional model of creating a "real" company with laboratories and full-time technical and business employees is simply too expensive an investment. It is capital inefficient with underutilization of infrastructure and human capital in companies inexorably limited to the development of a single lead compound, even while starting from a broad platform. When failure occurs in a small single-asset portfolio biotech company, project termination is not so easy. In practice, continuing to develop the drug on a "wing and a prayer" is often the path of least resistance taken. Thus, paradoxically, failure to kill projects in a timely manner is a more besetting sin in small biotech than in big pharma.

As a response in the last 2-3 years, old and new funds are turning to a much leaner, more "virtual" model of biotech investing, where the bulk of capital is used directly for drug development experiments, not administrative overhead. The plethora of contract service companies available to perform nowadays all components of the R&D process allows for a managerial group to avoid

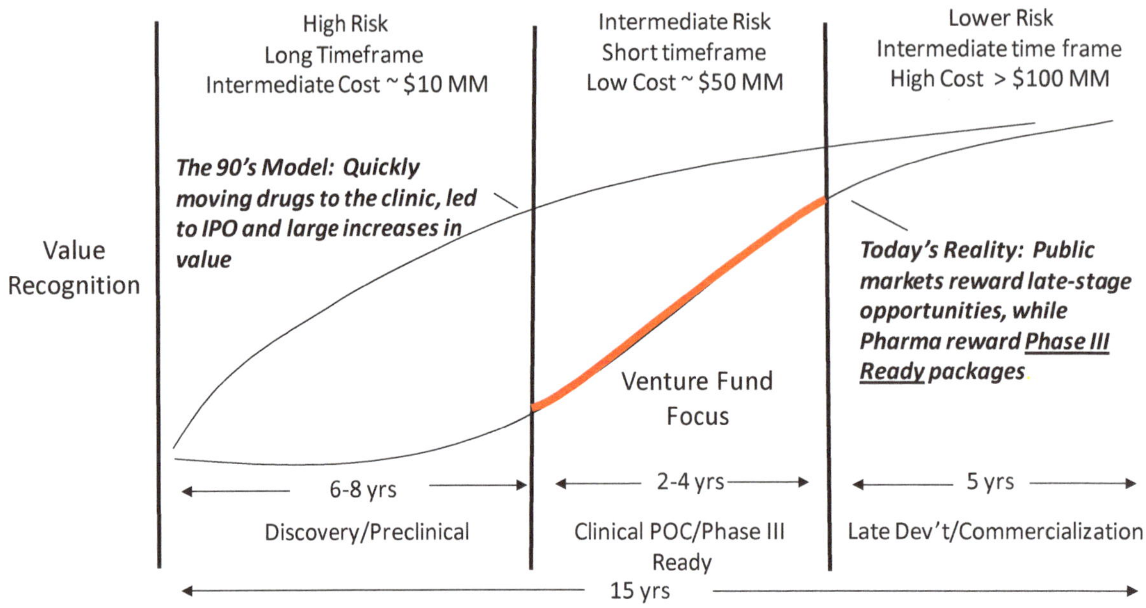

Figure 5: Adapting to the new exit environment: The "Back-end problem"
Adapted from Steve Mayer, Former CEO CoGenesys

much overhead. That managerial group may be the VC fund itself, or a development company (devco) subsidiary of the fund, or independently funded by LPs. This "virtual fund model" appears to work best if it is early stage, from the standpoint of risk-managing value accretion of more assets from late preclinical through clinical proof-of-concept with a finite capital pool, as opposed to fewer, later, clinical assets. Monetizing the assets is intrinsically easier for a virtual fund since there is no overhead or human capital for the pharma customer to have to purchase.

In the past two years, my partners and I have developed such a model for "virtual, early stage drug development" activity, called Druid BioVentures (DBV). DBV can be either a venture fund or a devco. Key components of the DBV model are (i) a coterie of very experienced, yet highly entrepreneurial, drug development managers, with whom I have had the good fortune to work over the years, to be the management "brains" for the early development portfolio, and (ii) off-shoring at least preclinical development and manufacturing activities to highly competent, regulatory agency-proven, CROs in countries like India and China, to maximize cost-effectiveness. Our asset opportunity trawl spans academic research centers and secondary projects of small biotech companies, as well as out-licensing opportunities from big pharma. In 2012, I am involved in another realization of this model — the well-funded Harrington Project originating at Case Western University and University Hospitals in Cleveland, Ohio.

CONCLUSIONS

We do indeed live in interesting times. I remain convinced that the global need for better drugs will only increase as the Western populations age, and the rest of the world matures financially, and, acquires deleterious Western lifestyle and dietary habits that promote degenerative disease. Even with the inexorable downward pressure on drug pricing, the pharmaceutical industry will remain profitable enough to attract significant capital. Drug R&D will become much more syndicated and globalized, and new sources of capital are emerging, especially in Asia. Until all the questions of biology and pathology arising from the genomics revolution are completely answered, drug research will always be a gamble for investors, who will nevertheless come to the table, and the future will be bright for the most knowledgeable and quickest-witted.

Project, product or company

Received: November 17 2011; revised January 6 2012

Arthur A. Boni
is John R. Thorne Chair of Entrepreneurship; Distinguished Career Professor; and Director, Donald H. Jones Center for Entrepreneurship, Tepper School of Business, Carnegie Mellon University.

ABSTRACT

This article addresses the issue that there are multiple options or paths to the market to be considered when developing the commercialization strategy to be employed for translation of a technology or invention into an innovation. We present a very simple screening methodology that may be applied to facilitate a quick but structured methodology for the entrepreneur to understand which option or options may be most viable to create, deliver and capture value. We use the concept of "project, product or company" as metaphors to categorize three commercialization pathways. Projects are best pursued with commercial partners via licensing arrangements. Products may be pursued using a research and development company business model. Company is intended to signify creation and growth of a lasting or durable organization intended to develop and bring multiple disruptive or sustaining innovations to market. Which path to the marketplace is appropriate, or even possible will depend on a number of factors that include understanding the magnitude of the value that is being created in the market, and the competitive landscape existing or likely to exist in that market. It is also necessary to determine whether the value captured by the business model that may be constructed is sufficient to balance the commercialization risks, while meeting the goals and objectives of the founders, investors and partners.

Journal of Commercial Biotechnology (2012) 18, 13–18. doi: 10.5912/jcb.508
Keywords: product; project; company; management

BACKGROUND

Many biotechnology innovations originate in academia since government funding often generates and advances science to the point where a proof of principle demonstration is achieved and commercial potential is apparent. It is necessary but not sufficient to have a proof of principle before advancing beyond the laboratory into the initial stages of defining a pathway to the market. Initial steps during the translational research stage can be taken within the university utilizing government funding to reduce the risk of technical failure and to understand more fully the clinical use. However, it is well recognized that if there is an intent to pursue a commercial pathway it will be necessary to "transfer the technology" from academia into a commercial domain. At this stage some private sector funding is often needed to move across the academic – commercial gap, which is the first "valley of death" experienced along the long commercial pathway, which will take many years and a substantial amount of private sector funding. This transition into the commercial domain can be effected through creation of a startup company intended to pursue the first commercial steps, or by licensing the technology/invention to an existing organization that may already be equipped to facilitate this step and subsequent commercialization. In either event the originators of the technology will continue to be involved with the process required to transfer the technology and the knowledge base acquired during the research phase that will be required to commercialize the technology. Our assumption here is that in either case it is prudent for the originators to analyze the pros and cons of the various options for the commercialization pathway.

The framework that we describe and utilize herein should be useful for all constituents involved in the potential new venture. There are two perspectives from which to consider innovations of a technological origin — technology push vs. market pull. In technology push, technology is developed first and foremost as a natural pursuit of research and development, and at some point in the future "market needs for the technology" are found. The technology breakthrough in fact comes before the market need is determined even though it may be anticipated in a very high level, non-specific sense. In

Correspondence: Arthur A. Boni, Ph. D., John R. Thorne Professor of Entrepreneurship, Tepper School of Business, Carnegie Mellon University, Pittsburgh, PA 15213 US. Email: boni@andrew.cmu.edu

market pull however, the market need is identified first, and then the technology is found that would enable a solution to fill this identified market need. This would be typified as a markets-first approach where an identified market need drives the creation of a specific solution. Examples of these two perspectives are plentiful. The laser is a classic example of technology push. Other examples may include the discovery of DNA in the biotechnology field as an enabling technology. Also consider stem cells, RNAi and antibodies. Markets however, often determine the evolution and use of these enabling (or disruptive) technologies once they are available to be considered as solutions for more specific identified needs. Treatment of specific diseases is a need in biotechnology, so scientists pursue development of vaccines, personalized medicines and diagnostics and many other "technologies" that are targeted at specific markets. So in most cases short of a disruptive fundamental discovery, most commercialized technologies lie between the two extremes. For example, the core or platform technology may lead to one early market entry application, and then the market determines the future product development evolution options (consider this as an emergent development strategy). The specific innovations are based on customer/user unmet needs.

We now consider another important issue that is often not well understood in the field of biotechnology and also in other industries driven by science and technology, i.e. robotics, nanotechnology, clean tech, etc. What is the value of the underlying technology in monetary terms? This issue is fundamental to the determination of the commercialization path that will be required to turn the invention into an innovation, as well as how to apportion value creation between the various participants required to bring a product to market. Simply put, innovation requires both an invention or unique underling technology and a business model to bring it to market. Several other sessions in the Biotechnology Entrepreneurship Boot Camp deal with the business model so suffice it to say here that the business model is defined as the rationale and arrangement of how an organization (and its partners) creates, delivers, and captures value. Some would argue that the intrinsic technology has little value without a business model to bring it to the market. Partnering or licensing can be seen as "renting the business model of the partner". So, often times the market values the technology at a level much less than the technologist can understand or is prepared to accept. The path or channel to the market, the understanding and processes needed to negotiate the clinical pathway, etc. most often garner a higher value in the commercial sense than the technology that often enables the business.

EVALUATING THE POTENTIAL INNOVATION

In this article and in the context of the Biotechnology Entrepreneurship Boot Camp we assume that the starting point for evaluation is the existence of some technology or platform that the founding team, and often the technology transfer office at a university wants to explore for commercial applications, potential, and identifying appropriate path(s) to the market. We start with the challenge to the reader (or Boot Camp participant) that it is really important for them to understand their objective(s) and goals underlying their desire to advance their technology into the commercial stage; what role they can or will play in that process; and, also to understand what is actually possible in the marketplace. Objectives may range from creating:

- A fully integrated pharma company (FIPCO)
- A research-intensive pharma company (RIPCO)
- A company that will enable the development of a product or platform

Or, another alternative may include enabling the advancement of the science or technology into clinical applications, but for the scientific founders to remain in academia or in the research environment while perhaps their doctoral students or post-doctoral students participate more directly in the commercial entity charged with commercialization. Inherent in this line of questioning is the recognition that individual ambitions and capabilities are important, as well as the risks taken by those who choose to move into the high-risk commercial environment, or invest in creating the infrastructure and the business model that will be required for commercialization. The latter point also includes building the team to found and execute the commercialization strategy. Team building and financing options can be very different for each of the alternatives listed above. Team building is specifically covered in another article published as part of this special issue and in the Boot Camp itself.

We suggest that opportunities should be screened initially using the "*quick screen*" proposed below to frame the discussion of alternatives. This initial step which can be done relatively quickly can then proceed with the inclusion of a more detailed analysis later since the screen most often raises a number of questions to be answered or issues to be resolved.

The quick screen is comprised of three questions to answer, all of which need to be answered affirmatively: 1) Is it an opportunity? 2) Can you win? 3) Is it worth it?

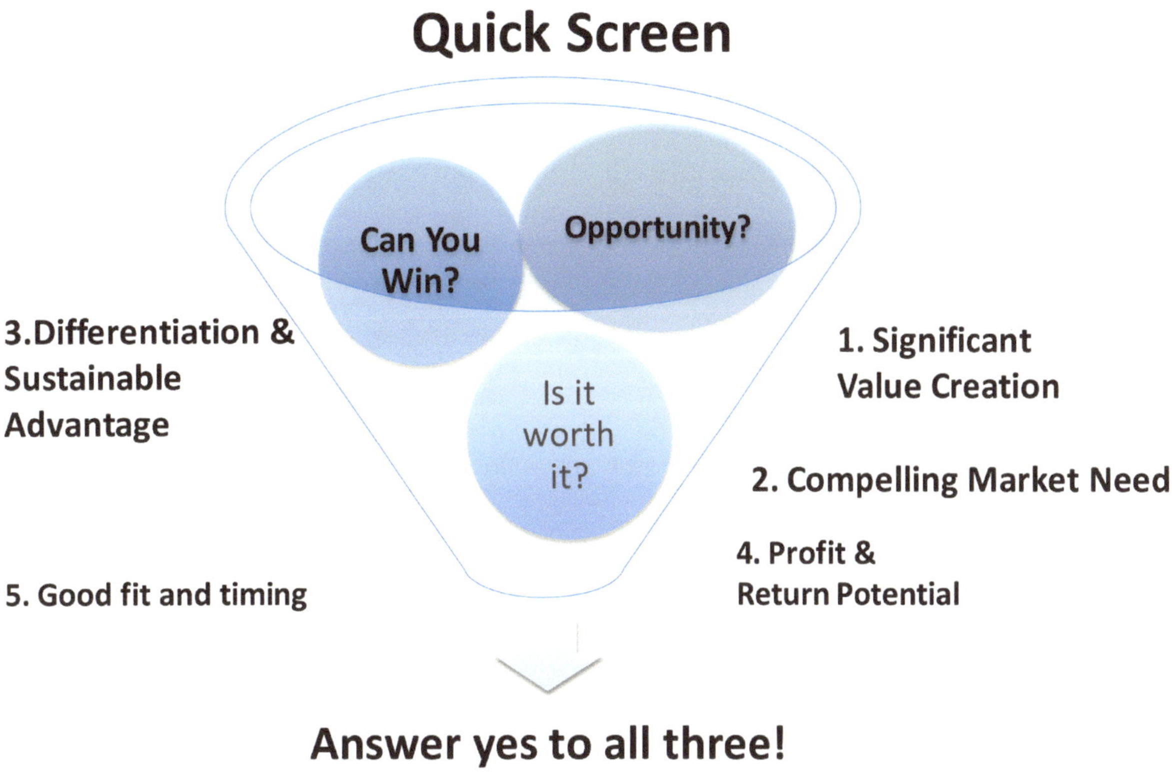

Figure 1: Quick screen and five anchors of a good opportunity

If you pass this screen, you have in effect determined that the opportunity could be molded to incorporate the *anchors of a good opportunity*. These "anchors" or pillars are the essential building blocks that comprise what is required for a successful venture. We have adapted this framework originally proposed by Timmons and Spinelli for new ventures[1]. However, we have added a fifth anchor (item 3 below) that in our opinion is essential for any successful technology-enabled venture (like biotech). This anchor deals with the importance of a differentiable solution with a sustained competitive advantage (via Intellectual Property and other elements of the business model). In our experience the presence of this pillar is important in any successful venture irrespective of the technology component. The five pillars are:

1. Creates or adds significant value to a customer or user
2. Solves a significant problem in a large and growing market
3. The opportunity can be differentiated and a sustained competitive advantage can be developed
4. The market has the potential for good margins and moneymaking characteristics
5. There is a good fit with the founders and management team at the time with a balance of risk and reward — and alignment of interest with all constituents including investors

The Quick Screen and Five Anchors of a Good Opportunity are illustrated in Figure 1. Given this framework we suggest that business, market and financing issues will drive the choice of commercialization option and financing strategy — along with the fit, timing, and risk profile noted above.

WHICH PATH TO THE MARKET? PROJECT, PRODUCT OR COMPANY?

Our proposed *Business Opportunity Screen* uses three dimensions, c.f. Figure 2:

1. Opportunity for unique value creation and strong customer/user need (Anchors 1 and 2 above)
2. Monetary or Economic Considerations (Anchor 4 above)
3. Differentiation and Competitive Advantage (Anchor 3 above)

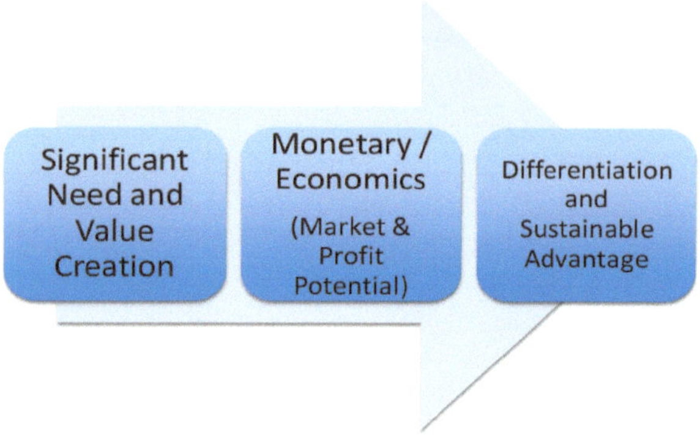

Figure 2: Business opportunity screen

The Business Opportunity Screen indicates whether the opportunity is best pursued as a project, product or company. The Monetary/Economic component may include the following considerations:

- How big can this business become?
- How much capital is needed and how long to reach positive cash flow?
- How much capital is needed and how long to reach profitability?
- How long to exit and what internal rate of return (IRR) or return on investment (ROI) is expected

In applying the Business Opportunity screen we recommend that use of an approach of ranking each component (opportunity, monetary/economic, and competitive advantage) as low, medium or high as a first pass. We recommend a fairly loose definition of low, medium and high. For example it would be helpful to think of market characteristics that would be of interest to an investor or partner. What is their time scale and magnitude of return on investment? Different perspectives will exist for VCs, angel investors and economic development groups. The same is true for government funding that supports commercialization. VCs will typically look for higher rate of return and shorter time to exit. Angels will also look for good returns but often have a longer time range to exit. Think in terms of the significance of the need in the context of its importance to the customer and his or her lack of satisfaction with current alternatives, i. e. how compelling is the need or "pain" in the market. The moneymaking or economics of the deal should be viewed in the context of "is it worth it" for the entrepreneur, partner or investor considering the level of risk and alternatives involved. Similar with competitive advantage, one must be sure that there are barriers to entry for other new entrants or competitors in addition to the IP that may exist or can be created.

One can always go back and add details as necessary for the decision making process, and also use a quantitative ranking methodology (risk adjusted) as necessary — the MBAs would probably do this. Alternately, think of this analysis as indicating areas of strength and weakness with the proposed pathway and to bring out the key questions that must be addressed to facilitate decisions and/or further work needed prior to making a decision. Therefore, we suggest that the screen be used as simply as possibly in a first pass. Then, based on the outcome, proceed by discussing the details of your analysis with some experts and customers to help with the evaluation by providing an outside perspective. In the Boot Camp we use a panel of experts to work with the participants on a real life "case" to apply this methodology. Since the participants and the panel have multidisciplinary backgrounds and perspectives, some very interesting discussions and debates occur. The same should happen for any such evaluation. Testing ideas and assumptions in the marketplace early and iteratively is strongly suggested along each stage of the commercialization path.

Characteristics of a Project — Most Often Described as a Good Licensing Option

Opportunity: Low due to small market, and the value to the customer is not compelling or significant; may be part of a solution but not a complete solution for the market.

Monetary: Low due to low money making potential (since other parts of the solution or its delivery will capture more of the total value in the market); also the technology may be so early stage that significant costs will need to be incurred to reduce the technology and/or clinical risks.

Competitive Advantage: Low since there may be other competitive solutions and companies, and while IP is possible the claims may be limited and the freedom to operate may require licenses or acquisition of complementary parts of the solution.

Recommendation or Approach

Consider licensing to an existing company; perhaps fund additional development via government or partner funding in a laboratory that already exists and with a scientific team in place; earn money via royalties and milestone fees.

Characteristics of a Product — Most Often Seen as a Development Stage Company Set Up to Commercialize a Product

Opportunity: Medium since market is apparent and value created is of interest, but is still not compelling or significant.

Monetary: Medium since while the technology may be further advanced there is still further development to be done before a significant "value for the technology" will be proven to exist.

Competitive Advantage: Medium for the above reasons, but once advanced to reduce technology, clinical and IP risks the solution would have barriers to entry in partnership with others with channel/market access.

Recommendation or Approach

In this case, it may be determined to move forward to further advance the technology down the clinical pathway to add value and reduce risk via a new company (NewCo). Funding can occur using non-dilutive sources first (government Small Business Innovation Research, for example). Complementary R&D can continue in a university and partnering with a larger biotech or pharma can occur downstream once a good value inflection point is reached. Once risk is reduced these "single product companies" could be acquired via the merger and acquisition process (M&A), which is often the result in biotech in any event, as pharma/larger biotech companies looks to fill product pipelines or to add technology platforms. Alternately, it may even possible to aggregate other companies with complementary product offerings to even add further value and to become a bigger player. If this latter path is followed, keep in mind that a more complete team may need to be built to advance the company further down the commercialization path (of course this will take more investment).

Characteristics of a Company — Built to Last with Multiple Products and Management Team to Carry Products Further thru the Regulatory Process

Opportunity: High since the need is significant, the solution compelling and the market large and growing (hundreds of millions of dollars annual sales possible).

Monetary: High profits and margins are possible and the product is advanced thru a value inflection point with the technology and also the clinical process has already been started so the return on investment potential is high of outside equity investors (partners and VCs).

Competitive Advantage: High since the solution is unique, differentiable and a strong IP position is or can be established. Better yet if a partnership is in place.

Recommendation or Approach

In this case, the founders must begin to develop and implement a plan to build and grow a sustainable organization with the potential for multiple products using the platform that is being created. Additionally, they will be required to advance the science/technology, the IP and to start the clinical demonstration process to prove safety and efficacy of the product and to validate the platform. They must also start the process of building a fundable team to balance the technology with business leadership. To accomplish all of these objectives it will be necessary to engage top tier investors to create a sustainable company with initial public offering (IPO) potential, or to have created enough value to attract an acquisition partner, thereby providing higher risk adjusted return to the investors via than the acquisition than would be achievable in the case of a Product Company. Very often in the M&A outcome, it is quite possible for the acquired company to continue operations (in part or in its entirety) as part of the acquiring company.

CONCLUSION

Keep in mind that in all of these cases the idea is to reduce risk incrementally in both technology and market while building a team that can move down the clinical path to market. At each stage the risk is reduced and the value of the "technology" and business opportunity is increased for all parties involved in the venture. See Figure 3, titled Innovation/Company Life Cycle that illustrates the principal of building value while reducing risk along two dimensions of product/market advancement (in the case of biomedical ventures market implies clinical progress) and building the team.

Capital efficiency is essential to keep in mind. For example, consider: leveraging the assets of partners or universities; using government or other non-dilutive sources of funding; keeping the management team very small (or almost virtual) for as long as possible.

Keep in mind that it is important to align the interests of the entrepreneur and the investor. Both need to be rewarded for their investment of resources (personal and time commitment) and financial assets. The return on investment is different for each party. The investor is often looking for a short-term financial gain. Venture capital investors have a 5 to 7 year time frame and look for a >10x return from each investment in their portfolio (to achieve an overall portfolio return of 3x). The entrepre-

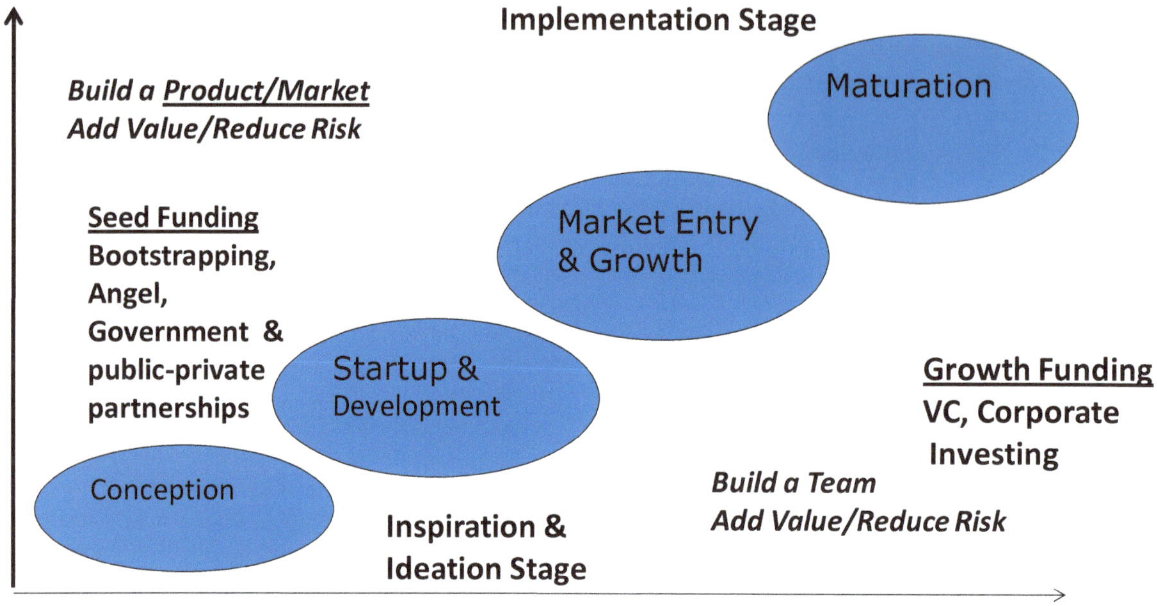

Figure 3: Innovation/company life cycle

neur looks for a financial return as well as seeing a successful commercialization as a result of their efforts. So, there is often a conflict between the short team returns that are financial in nature with the long term return that are often of a more qualitative or emotional nature. The need to align these interests is a key part of the success factor. Additionally there is a need for quality science, a quality team, and lots of preparation, as well as a little luck. At each stage of the company life cycle it is a good practice to make sure that each constituent of the venture (entrepreneur, investor, partner, employee) is meeting his or her respective goals and to understand that each has a legitimate position in advancing the venture towards commercial success.

REFERENCES

1. Timmons, Jeffrey A. and Spinelli, Steven, "New Venture Creation: Entrepreneurship for the 21st Century", McGraw-Hill Irwin (2007)

The basics of coverage, coding, and reimbursement for new medical devices and diagnostics: If you build it, will they buy it?

Received: November 15 2011; revised January 4 2012

Robert Wanerman
is a Senior Counsel at Epstein Becker & Green, P.C. in Washington, D.C. He provides regulatory and compliance counsel to a wide group of health care manufacturers, providers, and investors in coverage, coding, and reimbursement matters. He received an A.B. from The University of Chicago, a J.D. from New York Law School, and a M.P.H. from Columbia University.

ABSTRACT

The process of commercializing a new item or service in the U.S. health care market involves three distinct but necessary components: coverage, coding, and reimbursement. This article provides an overview of these processes and the challenges in successfully navigating the course and spotting the particular issues for individual items and services.

Journal of Commercial Biotechnology (2012) 18, 19–23. doi: 10.5912/jcb.514
Keywords: reimbursement; coverage; medical device; diagnostic

INTRODUCTION

For a medical device or diagnostic manufacturer, receiving approval or clearance from the Food and Drug Administration is a milestone event. However, manufacturers should consider the FDA process to be the end of the beginning of commercializing a product. This article will explain why a company should be planning for the coverage, coding, and reimbursement for that product and any related procedures as early as possible. Failing to do so can result in lost opportunities, significant additional costs, and dissatisfied providers and investors.

FUNDAMENTAL CONCEPTS

At the outset, several fundamental concepts must be nailed down to avoid costly errors once the company has developed an idea. Coverage refers to terms and conditions under which a private or public health plan will pay for an item or service.[1] For example, some new items and services may be covered initially by plans only after conventional treatments have been tried and failed; in other situations, the item or service might not be covered unless quantifiable diagnostic prerequisites have been met. Coverage is not guaranteed in all cases when the company receives FDA approval or clearance, and does not guarantee either a distinct code or a particular payment rate.[2] For this reason, waiting until the FDA approves or clears a new product to develop a commercialization strategy can result in a significant delay in bringing the product to market.[3]

Codes are unique identifiers for items or services that are billable or are related to billable items or services. Depending on the code set, they can define, among others, procedures (CPT codes), items (HCPCS Level II codes), revenue classification, and diagnoses (ICD-9 and 10 codes). They are used for claims processing and for

Correspondence: Robert Wanerman, Epstein Becker & Green, P.C., Washington, D.C., US. E-mail: rwanerman@ebglaw.com

1 *See, e.g., Hays v. Sebelius*, 589 F.3d 1229 (D.C. Cir. 2009).
2 *Goodman v. Sullivan*, 891 F.2d 449, 451 (2d Cir. 1989)(The Medicare program was not obligated to cover a diagnostic test based only on the FDA's approval of the test.)
3 CMS and FDA have started a pilot parallel review program under which a limited number of medical products would be considered by each agency simultaneously. 75 Fed. Reg. 57045 (2010) and 76 Fed. Reg. 62808 (2011).

research purposes, which can include how items or services are valued over time. As a result, it is important to define the codes that apply in a given situation. In certain settings, including many codes for radiology procedures, appropriate coding is made more complex due to the bifurcation of codes into a technical component and a professional component. Nevertheless, having a code is, by itself, not a guarantee of coverage or payment.[4]

The third basic concept is payment. This is a function of coverage and coding, and refers to the methodologies under which an item or service is reimbursed. Depending on the site of service and the individual plan policies, payment may be made for an item or service individually, or as part of a larger bundle. In addition, the amount of payment for the same service can vary among different sites of service due to factors such as indirect costs and overhead. For example, the Medicare program assumes that when the same service is performed in an ambulatory surgical center and a hospital outpatient department, the latter will have higher overhead and labor costs and sets a higher reimbursement rate for the hospital.[5]

The type and amount of payment a provider or supplier receives may be subject to additional constraints, including the use of relative value scales, payment caps or ceilings, and existing payment methods and rates for comparable items and services. To be sure, some companies and providers may be highly risk averse and will prefer to peg their coverage, coding, and reimbursement strategy to a comparable product. However, that strategy may have to be reevaluated if there is a new innovation that significantly alters the cost of the product or the time and effort needed by a provider or supplier who uses the new product. Conversely, some new technologies may substantially alter the overall cost of care so rapidly that bundled payment systems may not reflect the adoption of that new technology. In some limited cases, the Medicare program will approve payment for the new technology as an "add-on" cost or as a pass-though payment while data on the new technology is compiled and the bundled payment can be reviewed.[6]

RECOGNIZING THE POTENTIAL CHALLENGES AND PITFALLS

Developing a strategy for obtaining favorable coverage, coding, and reimbursement should start well before the product is launched in the market. In many cases, it is desirable to integrate these concerns into the design of a clinical trial involving the item or service; this allows for additional information beyond that required by the FDA to be gathered during the trial. This saves the time, effort, and cost of conducting a second trial to obtain the type of information that payors will demand in order to make a coverage determination.

IDENTIFYING THE STANDARD FOR COVERAGE

The basic standards for coverage are generally vague and give the payor wide latitude in approving coverage. The Medicare statute explains that items and services will be covered if they are "reasonable and necessary for the diagnosis or treatment of illness or injury" but does not explain the meaning of this term in more detail.[7] Some commercial plans rely on a general notion of medical necessity and accepted standards of practice.[8] In selected cases, items and services may be subject to a more rigorous evaluation process that examines improvements in outcomes and the relative benefits when compared to existing treatments.[9]

Since coverage is the threshold issue in developing a coverage, coding, and reimbursement strategy, the type of information and data that may be needed can turn on several important factors:

- Does the item or service fill an unmet clinical need?
- Who will benefit from the item or service; is it primarily one group (children or seniors), or is the benefit widely distributed?
- What is the anticipated site of service (physicians' offices, hospitals)?
- What is (are) the expected clinical outcome(s)?
- Are there items or services that are comparable, but inferior or superior (ex: screening or diagnostic tests with low sensitivity or specificity)?
- Will furnishing the item or service result in a facility fee and a professional fee?
- Is there a potential for coverage for "off-label" indications?
- What is the expected financial impact for the payer/consumer (ex: will adoption of

4 *See, e.g.*, Centers for Medicare and Medicaid Services, Innovator's Guide to Navigating Medicare, Version 2.0 (2010) at 17, available at: http://www.cms.gov/CouncilonTechInnov/01_overview.asp.

5 *Id.* at 31 – 55.

6 *Id.*

7 42 U.S.C. § 1395y(a)(1)(A).

8 *See, e.g.*, Regence Blue Cross and Blue Shield, Medical Policy Development and Review Process, available at http://blue.regence.com/trgmedpol/intro/

9 An example of this is the Blue Cross Blue Shield Association Technology Evaluation Center. *See* http://bcbs.com/blueresources/tec/

the new item or service raise or lower the aggregate cost of care)?

- Are there immediate or long-term benefits?

Another important consideration is how the item or service will be classified by providers and payors. Depending on the item or service and how it is to be furnished to a patient, it may be considered to be (1) an innovative breakthrough for patient health that fills a gap in treatment options; (2) something that replaces an existing item or service; or (3) adds or extends an existing technology or procedure.

Identifying the Type and Quality of Data to Support Coverage

For many payors, coverage is driven by the quantity and quality of data that the manufacturer or provider can produce to support coverage. The burden is usually placed on the applicant, and the type of data needed to support coverage can differ from the data submitted to the FDA to meet the safe and effective standard under the Federal Food, Drug, and Cosmetic Act. The shift toward evidence-based medical policy therefore relies on published, peer reviewed studies with sufficient statistical power to establish a given treatment hypothesis.[10] In order to satisfy this demand, manufacturers should rely more on randomized trials, other well-recognized study designs, and meta-analyses of published trials, and less on smaller studies or anecdotal information. In those cases where it may be impracticable or unethical to conduct a randomized trial due to recruitment problems, other designs such as case-control studies may be necessary.

In addition to data compiled to demonstrate the efficacy of an item or service, payors may demand more targeted data for certain sub-populations. This sort of challenge in gathering relevant data can be illustrated in a recent Medicare National Coverage Determination denying coverage for virtual colonoscopies using CT imaging.[11] Although many private health plans did offer limited coverage for CT colonographies, CMS focused its evidence review on how CT colonography compared to optical colonoscopy in the Medicare population, of whom the majority are over age 65. It concluded that the available evidence was deficient in part because the published studies contained only a few subjects who were eligible for Medicare, and because CMS could not determine if the results of CT colonography in younger subjects could be applied to the Medicare population.[12]

In selected cases, CMS may determine that a new technology has the potential for a significant benefit, but the quantity and quality of the data needed to make a final coverage determination is not yet available. Beginning in 2005, CMS developed an informal policy to approve conditional coverage and payment for these items and services while the additional clinical data on which a final decision can be made is being generated through a well-designed clinical trial. This policy, known as coverage with evidence development, can take one of two forms: (1) coverage with appropriateness determination, where additional clinical data is required to ensure that the item or service is being provided to appropriate beneficiaries according to established clinical criteria, and (2) coverage with study participation, where conditional coverage and reimbursement is approved, subject to the enrollment of affected individuals in a clinical trial that is expected to generate sufficient data in a clinical trial registry to allow CMS to make a final coverage determination.[13]

Manufacturers, providers, and suppliers must also be sensitive to the existing coverage environment. Depending on the item or service, coverage may be limited to specific clinical conditions, specific locations, or may be covered only if a first-line treatment has been tried but failed to produce results. The coverage environment can also be influenced by the interest of potential stakeholders or interested parties, which can include professional organizations, thought leaders in a particular specialty,

10 *See, e.g.*, D. Eddy, Evidence-Based Medicine: A Unified Approach, Health Affairs 24, no. 1 (2005): 9-17.

11 Centers for Medicare and Medicaid Services, Decision Memo for Screening Computed Tomography Colonography for Colorectal Cancer (2009), available at: https://www.cms.gov/medicare-coverage-database/details/nca-decision-memo.aspx?NCAId=220&ver=19&NcaName=Screening+Computed+Tomography+Colonography+(CTC)+for+Colorectal+Cancer&CoverageSelection=Both&ArticleType=All&PolicyType=Final&s=All&KeyWord=colonography&KeyWordLookUp=Title&KeyWordLookUp=Title&KeyWordLookUp=Title&KeyWordSearchType=And&KeyWordSearchType=And&KeyWordSearchType=And&bc=gAAAABAAIAAA& (Accessed November 8, 2011)

12 In addition, CMS was concerned that the data in the available studies found that the CT colonographies showed a lower sensitivity and specificity for polyps smaller than 6mm when compared to optical colonoscopy.

13 Centers for Medicare and Medicaid Services, Coverage with Evidence Development Solicitation (Nov. 7, 2011), *available at*: https://www.cms.gov/medicare-coverage-database/details/medicare-coverage-document-details.aspx?MCDId=8&McdName=National+Coverage+Determinations+with+Data+Collection+as+a+Condition+of+Coverage%3a+Coverage+with+Evidence+Development&mcdtypename=Guidance+Documents&MCDIndexType=1&bc=BAAIAAAAAAAA&. CMS is considering changes to its current policies, but has yet to publish any proposals for comment.

advocacy groups, and the providers or suppliers who will be using a product or performing a procedure using that product.

Another factor that potentially impacts coverage decisions is the growth of concerns for the comparative effectiveness of treatments for a given condition, as opposed to the efficacy of a treatment for initial approval or clearance by the FDA. To date, these initiatives have focused on a mix of outcome and cost variables, and have taken several forms including government-funded analyses through entities such as the CMS, the Agency for Healthcare Research and Quality, or the U.S. Preventive Services Task Force, and through privately funded research, such as the Blue Cross Blue Shield Technology Evaluation Center, or initiatives developed by the Kaiser Permanente Division of Research. These decisions can affect both new items and procedures, and those that have been in use for some time.[14]

With the enactment of the Patient Protection and Affordable Care Act in 2010, the scope and potential impact of comparative effectiveness research was significantly expanded through the establishment of a nonprofit Patient-Centered Outcomes Research Institute ("PCORI"), which will receive up to $150 million annually in federal funding.[15] The institute will be governed by a 19-member board drawn from government, industry, health professions, and researchers, and is charged with identifying national priorities for research, taking into account disease incidence, prevalence, burden in the U.S. (emphasizing chronic conditions), gaps in clinical outcomes evidence, practice variations, impact on national health expenditures, and "the potential for new evidence to improve patient health, well-being, and the quality of care..."[16] PCORI's research must take into account effectiveness within subpopulations (racial, ethnic, gender, age, comorbidities, genetic and molecular subtypes, and quality of life preferences). Nevertheless, due to concerns about rationing of care, Congress included specific limitations on the use of PCORI's research. It cannot mandate coverage or reimbursement, and cannot make recommendations based on "dollars per-quality adjusted life year" as a criterion. Moreover, although its findings can be adopted by CMS for Medicare coverage purposes, it can only do so through a process that includes public notice and comment, and any such decisions cannot override current Medicare coverage determinations or be the exclusive basis for a change in Medicare policy.[17]

14 *See, e.g.,* G. Jacobson, CRS Report for Congress: Comparative Clinical Effectiveness and Cost-Effectiveness Research: Background, History, and Overview (2007).

15 Pub. L. No. 111-148, §§ 6301 and 10602.

16 *Id.*

17 *Id.* The statute is silent on the use of any research or recommendations by a private health plan.

Positioning the Product for Favorable Reimbursement

The third component of a commercialization strategy is obtaining favorable reimbursement for the item or service. Once the manufacturer or provider can confirm that the item or service will be covered, the next set of questions to be answered can place the item or service in the appropriate context. This can turn on five important points: (1) the code(s) that currently exist; (2) the reimbursement methodology for the item or service in a particular setting; (3) the range of reimbursement for those codes; (4) whether or not the total reimbursement is acceptable; and (5) whether or not there is persuasive evidence to justify enhanced reimbursement.

Depending on where and how the item or service will be delivered, favorable reimbursement can have a different meaning. For example, many hospitals are now paid a fixed sum for inpatient or outpatient services, which are known as Diagnosis Related Groups for inpatient services, and Ambulatory Payment Classifications for outpatient services.[18] In either setting, the manufacturers and other suppliers must be sensitive to these bundled payments. In the hospital example, if adopting the new product or technology raises its costs, it also decreases the hospital's overall profitability unless the net impact lowers the overall cost of care or allows patients to be treated more efficiently. For some manufacturers, establishing a coverage and reimbursement strategy may be more complex if the same item can be used in multiple settings with different reimbursement methodologies.

For some complex situations, coordinating separate reimbursement components may be a crucial part of the success of a new technology. For example, if a new item or technology such as a joint replacement can be furnished safely only in a hospital, ambulatory surgical center, or other facility, and requires the services of a physician, then the adoption of that new item or technology will depend on the reimbursement for *both* the facility and the physician.

Spotting the Potential Coverage, Coding, and Reimbursement Issues for Payors and Providers

After marshalling the clinical evidence needed to persuade payors to cover a new item or service, and compiling the financial data to support reasonable reimbursement for the item or service, the existing environment from the perspective of payors and providers may still have some traps for the unprepared. First, some payors

18 The same bundling concept has also been adopted for other types of care, including skilled nursing services, home health care, and ambulatory surgical services.

may be unwilling to cover a new item or service if it increases costs significantly without producing improved outcomes in the short run or reduces the long-term costs of an episode of care. A variant of this approach that may arise when dealing with some government programs is a budget neutrality constraint that may deter the adoption of expensive new technologies.

A special example of this problem can occur when a manufacturer of a new device obtains clearance from the FDA under Section 510(k) of the Federal Food, Drug, and Cosmetic Act after demonstrating that its product is substantially equivalent to a legally marketed device. The benefit of this approach is that less time is needed for FDA review. The potential risk is that payors will treat the new product exactly the same as the predicate device, and apply the same coverage and reimbursement criteria. Accordingly, manufacturers should have additional clinical evidence for health plans and other payors that demonstrate improved outcomes or different costs based on coverage and reimbursement criteria, and that is distinguishable from the standards used by the FDA.

Another related potential pitfall for some new products is the application of payor policies setting reimbursement for a class of items at the rate paid for the "least costly alternative." In the Medicare program, this policy exists only as a non-binding interpretation in a program manual. While it has been applied to certain items of durable medical equipment, CMS was blocked in its attempt to apply the policy to certain drugs when the policy conflicted with express statutory and regulatory language establishing a reimbursement formula.[19]

A new item or service may also need a new code in order to accurately reflect its characteristics and to differentiate it from other similar items and services. Individual payors or health plans may make coding determinations, or may provide confirmation of the correct code to use. If a new code is needed, the factors that the entities that control the code sets, such as the CPT Editorial Panel or the HCPCS Workgroup, will rely on include technological improvement, clinical improvement, and the need for higher and more complex resources to render a service. The fact that a manufacturer or provider may be seeking enhanced reimbursement for that item or service will carry little weight. An additional factor is the time and effort needed to obtain a new code; obtaining a new CPT code follows a well-defined process that takes up to two years, whereas the HCPCS process is more discretionary but can be completed within one year. Therefore, interested parties should expect that coding and possibly reimbursement will lag behind technological development.

Third, providers and suppliers that are subject to bundled or capitated reimbursement rates may be reluctant to adopt a new technology that increases their costs (and shrinks or eliminates their profit). In the case of smaller providers or physician groups, adoption of a new technology that requires a significant capital expense may be a deterrent unless there is generous coverage and reimbursement; if the pace of innovation is rapid, they may delay adopting a new technology to avoid the additional costs of new equipment and supplies, along with the lag period that may be needed to become proficient with the new technology.

In addition to these potential problems, even well-designed coverage and reimbursement strategies cannot control the possibility of opposition from entities that may stand to lose from the adoption of a new item or service. For example, hospitals may oppose the expansion of procedures that can be performed in ambulatory surgical centers, or physician specialty groups may oppose the introduction of new items or services that can be performed by mid-level practitioners.

MAKING THE PROCESS WORK

A successful commercialization strategy for a new item or service in the health care system involves assembling the coverage, coding, and reimbursement pieces much like a jigsaw puzzle: it takes patience, perseverance, excellent information, help from others, and good timing. Manufacturers, providers, and payors should all recognize that the challenges will be different in almost every case because of many of the variables discussed in this article. Assembling some parts of the puzzle will be easier than others, and some items may be outside of the control of an interested party. Nevertheless, planning for those contingencies as early in the process of commercialization as possible can have a significant return when it is most needed.

19 *See Hays, supra*, fn. 1.

Transition from the lab to the clinic — Regulatory considerations

Received: November 11 2011; revised February 6 2012

James G. Kenimer
founded BCG in 1993 after thirteen years with the FDA where his last position was Deputy Director of the Division of Vaccines and Related Products Applications at CBER. He served as President and CEO of BCG until his retirement in October, 2011. He is currently Chairman of the Board of BCG. He has extensive experience in the regulatory approval process for biologic products, with particular interests in vaccine and therapeutic proteins. Dr. Kenimer earned a PhD in biochemistry from Emory University and was a post-doctoral fellow at UCLA, Georgetown University and the National Institutes of Health prior to joining the FDA in 1980.

Jim Ackland
has over 38 years experience in the manufacture, quality control, development and international regulatory requirements for biopharmaceutical products. This includes senior management positions in vaccine manufacture, quality assurance and research and development. From 1990 to 1999 he was Head of Regulatory Affairs for Vaccine Products at CSL Limited (Australia). From 1999 to 2005 he was Vice President of Biologics Consulting Group (USA). Jim is currently CEO of Global BioSolutions, an Australian company based in Melbourne Australia, which provides regulatory and product development advice and assistance to biotech companies.

Journal of Commercial Biotechnology (2012) 18, 24–30. doi: 10.5912/jcb.506
Keywords: regulatory approval, FDA, law, regulation

INTRODUCTION

Companies trying to transition what seemed like a good idea for a new pharmaceutical product in the laboratory into clinical development face a myriad of regulations and guidelines. The empirical trial and error based science has to give way to the highly regulated structured environment of pharmaceutical development. Regulations prescribe standards for manufacturing, safety testing and clinical development. In addition prior to commencing clinical development regulatory approval is required to assess if there is adequate information on the product to proceed.

Understanding the regulatory requirements for each aspect of development is critical, as is having a well developed plan taking into account the regulatory expectations. Companies often find it useful to conduct a regulatory gap analysis which covers all aspects of the development and to produce a comprehensive Product Development Plan. Having a multi-discipline product development team including preclinical, manufacturing, clinical and project management is an essential element to successful development. Effective communication with regulatory agencies can help avoid costly delays due to the right information not being available at key time points. In addition management support is needed for this team effort and success is highly dependent on upper management support and availability of appropriate resources. As products move through development, the regulatory scrutiny increases.

Before considering the specific regulatory requirements and expectations relating to your product it is informative to understand who regulates and why.

Correspondence: James Kenimer, Biologics Consulting Group, Inc., Alexandria, VA, US. E-mail: chairman@bcg-usa.com

US LAWS, REGULATIONS AND GUIDELINES

In the United States, Drugs, Biologics and Medical Devices are regulated by the U.S. Food and Drug Administration (FDA). Drugs and medical devices are regulated under authority of the Food, Drugs and Cosmetics (FDC) Act and biologic products are regulated under authority of the Public Health Services (PHS) Act.
Currently the FDA is organized in seven Centers:

a. Center for Drug Evaluation and Research (CDER)
b. Center for Biologics Evaluation and Research (CBER)
c. Center for Devices and Radiological Health (CDRH)

d. Center for Food Safety and Applied Nutrition (CFSAN)
e. Center for Tobacco Products (CTP)
f. Center for Veterinary Medicine (CVM)
g. Center for Tobacco Products (CTP)

The evolution of the regulation of drugs, medical devices and biologic products by the FDA makes for interesting reading. Selected important dates are summarized below (more details are available on the FDA website at http://www.fda.gov/AboutFDA/WhatWeDo/History/default.htm).

- 1902 – Biologics Control Act – Passed by Congress in response to an incident in St. Louis, Missouri in which a diphtheria antisera — later found to be contaminated with tetanus toxin — led to the death of 13 children. The Act was focused on the purity and safety of serums, vaccines and similar products and formed the initial legal basis for the regulation of biologic products in the US. Authority for enforcement was assigned to the Hygienic Laboratory of the Public Health and Marine Hospital Service (the predecessor of the current Center for Biologics Evaluation and Research).

- 1906 – Food and Drugs Act and the Meat Inspection Act – Passed by Congress following public disclosures about the unsanitary conditions in meat-packing plants (Upton Sinclair's *The Jungle* was one of the most influential), the use of poisonous preservatives and dyes in foods, and cure-all claims for dangerous patent medicines. This Act prohibited the interstate commerce of misbranded and adulterated foods, drinks and drugs. Authority for enforcement was assigned to the Bureau of Chemistry, Department of Agriculture (the predecessor of the current Food and Drug Administration).

- 1927 – Bureau of Chemistry reorganized with regulatory functions assigned to a new Food, Drug and Insecticide Administration (later shortened to the Food and Drug Administration – FDA). The FDA remained in the Department of Agriculture. [Note that biologic products were not included in those products regulated by the FDA – they remained under the regulation of the Hygienic Laboratory in the Public Heath Service].

- 1938 – The Federal Food, Drug and Cosmetic (FDC) Act – Updated the 1906 Food and Drugs Act to extend control to cosmetics and therapeutic devices and to authorize factory inspections (factory inspection of biologic products already allowed under the 1902 Biologics Control Act). This Act required that new drugs be shown to be safe prior to marketing (the 1906 law only required that they be properly labeled).

- 1944 – The 1902 Biologics Control Act was expanded and incorporated into the new Public Health Services Act (the current law under which biologic products are regulated in the US). The authority for regulation of biologic products was transferred to the NIH Microbiological Institute (the first NIH Institute). Note that biologic products were still not regulated by the FDA.

- 1955 – The "Cutter Incident" – Congress responded to another tragedy in response to the administration of improperly inactivated polio vaccine to children resulting in 260 cases of polio in vaccine recipients. The authority for regulation of biologic products was transferred to a new entity within the NIH – the Division of Biologics Control (biologic products still not regulated by the FDA).

- 1962 – Kefauver-Harris Drug Amendments – required drug manufacturers (including manufacturers of biologic products) to demonstrate efficacy of products as a condition of approval. Prior to this, only safety and purity were required.

- 1972 – Regulation of biologic products transferred from NIH to the FDA (the resultant Bureau of Biologics remained on the NIH campus)

- 1976 – Medical Device Amendments to the FDC Act passed to assure the safety and effectiveness of medical devices.

Established the pre-market approval processes for medical devices.

- 1983 – Orphan Drug Act – Enabled FDA to promote the research and marketing on drugs and biologics for the treatment of rare (orphan) diseases.

- 1984 – The Drug Price Competition and Patent Term Restoration Act – Expedited the availability of less costly generic drugs (but not biologic products) by permitting FDA to approve applications to market generic versions of brand-name drugs without having to repeat the studies done to prove safety and effectiveness.

- 1992 – Prescription Drug User Fee Act – Requires drug and biologic product manufacturers to pay fees for license applications (New Drug Applications (NDAs) and Biologic License Applications(BLAs)) and supplements. The Act required FDA to use the funds to hire more reviewers.

- 2002 – Medical Device User Fee and Modernization Act – Extended the user fee payment requirements to sponsors of medical device applications.

- 2004 – Project Bioshield Act – Authorizes FDA to expedite review procedures to enable rapid distribution of treatments as countermeasures to chemical, biological and nuclear agents that may be used in terrorist attacks.

- 2007 - FDA Amendments Act (FDAAA) of 2007 – Numerous amendments to the FDC Act including the reauthorization and expansion of the Prescription Drug User Fee Act (PDUFA) and the Medical Device User Fee and Modernization Act (MDUFMA). Two other important laws were reauthorized: the Best Pharmaceuticals for Children Act (BPCA) and the Pediatric Research Equity Act (PREA). Both of these are designed to encourage more research into, and more development of, treatments for children.

- 2009 – FDA's Center for Tobacco Products established.

- 2010 - The Biologics Price Competition and Innovation Act (BPCI Act) enacted as part of the Affordable Care Act. The BPCI Act created an abbreviated licensure pathway for biological products (biosimilars) which is similar in concept to the generic drug approval pathway for drugs established in 1984 (see above).

In response to laws passed by the U.S. Congress (see examples above) the FDA develops and publishes regulations, following procedures as required by the Administrative Procedure Act, such that the intent of the law is translated into specific rules to be followed by sponsors as they develop their drug, medical device, or biologic product. The process of developing these regulations/rules typically involves a "notice and comment rulemaking" process for seeking public comment on the proposed regulations before issuing the final regulation. These regulations are legally binding on both the FDA and the public.

All U.S. regulations are published in the Code of Federal Regulations (CFR). All regulations pertaining to FDA issues are contained in Title 21 of the CFR (referenced as 21 CFR). All human drug regulations are contained in sections 300-399 of 21 CFR; all biologic product regulation in sections 600-699 of 21 CFR; and all medical devices regulations in sections 800-899 of 21 CFR. All the regulations are easily accessed through the internet.

In addition to the published regulations, the FDA also develops and publishes Guidance Documents, following procedures required in its "Good Guidance Practice" regulation. These Guidance Documents describe the Agency's current thinking on a particular regulatory issue; however they do not have the force of law and are not binding on either the FDA or the public. Although there are many guidance documents they are often generic in nature and need to be interpreted in relation to the specific product or situation. The guidelines can be found on the US FDA website www.fda.gov.

INTERNATIONAL REGULATIONS

Each country has its own laws, regulations and guidelines which control import, clinical development and marketing of pharmaceuticals and medical devices and country specific approvals are required. For example, in Europe, the European Medicines Agency is the peak regulatory body and approves most new pharmaceuticals although individual country regulatory agencies, such as the Medicines and Healthcare products Regulatory Agency (MHRA) in the UK and the Medical Products Agency (MPA) in Sweden approve clinical studies.

In addition to country or region specific guidelines, the World Health Organization publishes guidelines and standards on pharmaceutical products which may be used by international regulatory agencies.

To help deal with this myriad of guidance, which is sometimes inconsistent, the US, Europe and Japan representing the major pharmaceutical markets have worked to harmonize many of the more general guidance documents. This process is called the International Congress of Harmonisation (ICH) and involves regulatory agency and pharmaceutical company representatives working together to address inconsistencies. ICH guidelines have helped harmonize many aspects of pharmaceutical development and in particular the preclinical toxicology requirements which had been quite disparate between the three regions for some product types. Despite this harmonization effort there are still country/region specific requirements.

PHASES OF PRODUCT DEVELOPMENT

Product development consists of fairly well defined phases.

- Discovery basic research
- Process and analytical development
- Preclinical animal studies
- Phase 1 clinical studies
- Phase 2 clinical studies
- Phase 3 clinical studies
- Product approval/licensure
- Post marketing studies (Phase 4)

The basic research phase usually involves *in vitro* and/or *in vivo* models to generate a rationale for the development of the product, proposed mechanism of action and selection of a lead product.

Once the product concept is chosen a suitable manufacturing process needs to be developed which can be scaled to commercial manufacturing. It is worth noting that some processes and reagents used in research laboratories are not particularly compatible with commercial pharmaceutical manufacture. In addition, analytical tests are required which will characterize the product in terms of identity, purity, impurities and quantity/potency. There is an expectation from regulatory agencies that analytical tests are adequately qualified or validated. During this phase, early evidence of product stability should be demonstrated and stability indicating assays developed.

Although there is no regulatory oversight during the research and early preclinical phase, a working knowledge of the regulatory expectations based on the guidelines and regulations can avoid costly delays due to inadequate characterization of the product, use of an inappropriate process or inadequate purity of the product. A review of the regulatory guidelines in relation to the requirements for manufacture and testing is essential during this early phase of development.

Preclinical animal studies are required to characterize the potential toxicities associated with the drug product and to define a clinical starting dose. Although there are US and international guidelines which prescribe certain tests, for many biological products there are nuances

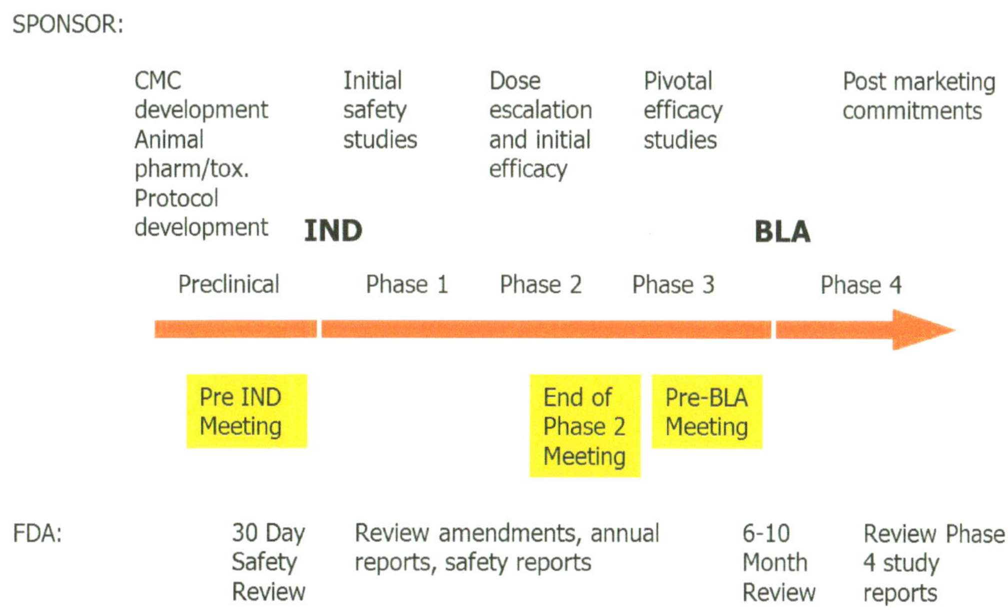

Figure 1: Phases of product development

in terms of choice of animal species and dose selection. Expert opinion on the types and design of studies should be sought.

A typical preclinical package for a new chemical entity to support initial clinical development would include the following:

- Single and repeat dose toxicology
  - Minimum 14 days toxicology, rodent and non-rodent species
- "Special" toxicology
  - Local tolerance, sensitisation
- Mutagenicity
  - Bacterial mutation *(in vitro)*, chromosomal aberration **or** lymphoma assay
- Safety pharmacology
  - Major body system effects — cardiovascular, respiratory, CNS
- Efficacy pharmacology
  - Demonstrate mode of action from *in vitro / in vivo* models of activity
- Absorption, Distribution, Metabolism, Excretion (ADME)
  - Compare metabolic fate *in vitro*
  - Demonstrate lack of accumulation

The preclinical requirements depend on the type of product and the indication. For example many biological products which may consist of naturally occurring amino acids and nucleic acids usually require only a single species toxicity study and are exempt from mutagenicity and ADME studies.

Figure 2: Regulatory hurdles

REGULATORY HURDLES

At the conclusion of the preclinical phase the transition into the clinic can occur. This involves the submission of a clinical trial application (Investigational New Drug (IND) in the US) to the relevant regulatory agency and a regulatory review of the information relating to the manufacture and testing of the product. Prior to submission of the IND the FDA offers companies the opportunity to discuss the information requirements at a pre-IND meeting. This is always extremely useful in assessing what information is required to permit a product to enter clinical development. The meeting procedures are described in an FDA guidance document. In a similar manner, many countries offer the ability to meet prior to clinical development or during clinical development to discuss requirements. In Europe this can be at the country regulatory agency or at the central European Medicines Agency. These meetings are extremely important in providing guidance on the type of preclinical studies and the expectations of regulatory agencies during development.

Later in development further meetings with FDA and other international agencies are encouraged. These meetings are extremely important in terms of the design and endpoints of late stage clinical trials.

The transition of a pharmaceutical product from the research lab to the clinic (translational development) is a highly regulated process with numerous steps where FDA regulations and guidance must be rigorously applied. The consequences of non-compliance can be severe.

FDA "Good Practice Regulations" apply at every step of product development. At the preclinical pharmacology/toxicology animal testing stage, important studies are required to be conducted under the Good Laboratory Practice (GLP) regulations [21CFR58]. Manufacture of products intended for use in humans are required to meet the escalating requirements of the Good Manufacturing Practice (GMP) regulations [21CFR210-211 for Drugs; 21CFR600 Subpart B for Biologics] , which significantly increase in complexity as the development transitions from Phase 1 through Phase 2 to the pivotal Phase 3 clinical trials. The clinical trials themselves must be conducted in compliance with the Good Clinical Practice (GCP) regulations.

Prior to conducting clinical studies in humans, the FDA requires that specific information about the product and the proposed clinical trails be submitted for review and approval by FDA reviewers. For drugs and biologics the required regulatory submission is termed an Investigational New Drug (IND) submission. Details about the information required to be submitted in the IND is published in 21 CFR Part 312 and further clarified in various FDA guidance documents. The IND is essentially a request to be exempt from U.S. regulations which prohibit the administration of non-approved products to humans. The equivalent regulatory submission for a medical device is termed an Investigational Device Exemption (IDE) and the regulations governing an IDE are published in 21 CFR Part 812.

The product development regulatory goals can be defined as:

1. Develop a reproducible process that can yield a consistent product and that can be run under GMP regulations
2. Develop analytical procedures that can reliably measure product parameters including stability, and that can demonstrate product comparability following manufacturing/facility/equipment changes.
3. Develop animal models that can demonstrate proof-of-concept and safety.
4. Demonstrate safety and efficacy in human clinical trials

CLINICAL DEVELOPMENT — REGULATORY CONSIDERATION

The generally accepted progression of clinical development involves three phases, and the regulatory requirements and expectations significantly escalate as a sponsor advances through these stages.

Phase 1: Phase 1 trials involve the initial introduction of the test product into humans. Demonstration of safety is the primary objective; therefore the trials generally include small numbers of subjects who are closely monitored for adverse events. Depending on the product and the intended indication, Phase 1 trials are sometimes conducted in healthy volunteers to more easily evaluate the appearance of any adverse events.

Phase 2: Phase 2 trials are generally described as hypothesis-generating trials, in that they are designed to provide data with respect to both safety and efficacy in an appropriate patient population and to generate a hypothesis of the proposed mechanism of action of the treatment. The objectives are to optimize dose, route of administration, treatment regimen, patient population and clinical endpoints. Well designed and properly conducted Phase 2 trials are extremely important to a successful product development — however many products fail later stage pivotal trials because insufficient data was collected during Phase 2 to adequately design the pivotal trial(s).

Phase 3: Phase 3 trials are the final pivotal hypothesis-verifying trials intended to supply the data which will be used as the basis for the regulatory approval submission (the New Drug Application (NDA) for a drug or the Biologics License Application (BLA) for a biologic). They ideally are randomized and placebo-controlled and large enough to provide sufficient statistical power for proof of efficacy. The objectives are to generate sufficient data to support regulatory approval; to establish a risk-benefit ratio and to support labeling claims.

Poor regulatory strategy and lack of attention to regulatory expectations can and does lead to significant delays and failures in product development. Inadequate animal studies, inadequate bench testing, poor product characterization and insufficient validation have all led to FDA stopping a clinical trial from proceeding. The regulatory mechanism for this is called a clinical hold.

Clinical holds can have a devastating effect on the company with significant rework and delays sometimes forcing companies to close, seek additional funding, and delay proposed exit strategies.

PRODUCT DEVELOPMENT PLANNING

Product planning is critical to any organization, and a well-conceived and comprehensive Product Development Plan (PDP) can provide a detailed assessment of your product and the most effective pathway to licensure/approval.

The PDP is a "roadmap" for your product's development. It should be concise, product-focused strategic document laying out the path to licensure/approval and include a detailed analysis of your product status and developmental requirements, including the four primary aspects of product development: Manufacturing, Preclinical, Regulatory and Clinical Development. It is an integrated stand-alone document tying the four main areas of product development with budgets, tasks and timelines through Phase 1 or beyond.

A well developed PDP is crucial at every stage of development, particularly at the outset. It can provide a concise detailed analysis of your product and the roadmap to market with clear developmental objectives and crucial milestones. It can present a single (or multiple, if desired) focused regulatory strategy for presenting your product to the FDA. It should include strategies for dealing with potential risks in the product development process and lay out accurate and realistic budgets and timelines through clinical development.

The typical PDP covers the following:

- Background and Product Assessment
- Manufacturing Development Plan
- Preclinical Development Plan
- Clinical Development Plan
- Regulatory Development
- Project Management
- Budget
- Timelines
- Risk and mitigation strategies

Regulatory compliance is critical to success, and keep in mind that if the FDA doesn't approve it you can't test it in humans and more importantly you can't sell it. Achieving regulatory compliance is hard work and requires a significant dedication of resources by product development specialists who have expertise with your product type. A rigorous PDP with consideration of the regulatory expectations will provide a roadmap to speedy approval.

Building teams in entrepreneurial companies

Received: November 11 2011; revised February 7 2012

Arthur A. Boni
is John R. Thorne Chair of Entrepreneurship; Distinguished Career Professor; and Director, Donald H. Jones Center for Entrepreneurship, Tepper School of Business, Carnegie Mellon University.

Laurie Weingart
is Carnegie Bosch Professor of Organizational Behavior and Theory at the Tepper School of Business, Carnegie Mellon University.

ABSTRACT

This article focuses on the essentials of building effective, collaborative, team-based organizations. Our target audience is the entrepreneurs and innovators who found and build knowledge-based organizations, especially entrepreneurs in the biotechnology and biomedical fields who are technologically driven, but who also have special constraints typified by these industries, e.g. long life cycles, highly capital and risk intensive, and also highly regulated. We address best practices for building entrepreneurial companies from two perspectives: 1) "experiential learning" gathered over years of experience in building and growing entrepreneurial organizations; and, 2) "academic learning" on building effective teams based on selected academic or scholarly literature. Our goal is to provide a perspective that blends real-world lessons filtered through a more scholarly approach based on case literature and other research-based studies. The material summarized herein is presented as a learning module to the participants in the Biotechnology Entrepreneurship Boot Camp. The pedagogical approach taken is to present the background material and perspective contained herein to provide a blended experiential and academic perspective. As a counter-point, we follow with a moderated panel discussion around these and other topics from the real-world perspective of biotechnology entrepreneurs. The panel consists of the key members of a real biotechnology company, different for every Boot Camp, consisting of key C-level officers and founders and a venture capitalist that funded the company. Thus the "theory and practice" of building a biotechnology company comes together in the context of a real-time case with audience interaction. We summarize briefly in a concluding section some key lessons learned from numerous panel discussions, which have been gathered over the years.

Journal of Commercial Biotechnology (2012) 18, 31–37. doi: 10.5912/jcb.507
Keywords: team; management; performance; start-up

INTRODUCTION

The onset of company formation starts with the vision of the founders and the articulation of the culture that they want to build into the "DNA" of the startup organization. The goal is to build and then sustain that vision and culture as the company grows through its life cycle. While each company is different in regard to its culture and mission, the challenge faced by the founders can be reduced to the following ingredients articulated by Boni in a review of the book written by the second CEO of Amgen, Gordon Binder[1]:

- Build a talented and balanced management team in a culture that incorporates an interdisciplinary, team-based, collaborative approach with leadership throughout
- Encourage and reward performance
- Organize around autonomy and innovation
- Tolerate risk and learn from failure

All of us can learn a few lessons from Amgen, which is arguably one of the most successful biotechnology companies in the relatively short history of the industry. We suggest that in building a management team there

Correspondence: Arthur A. Boni, Ph. D., John R. Thorne Professor of Entrepreneurship, Tepper School of Business, Carnegie Mellon University, Pittsburgh, PA 15213 US. Email: boni@andrew.cmu.edu

are some best practices that have proven to be successful over the years. First and foremost is the challenge and principal objective to *build an entrepreneurial culture* that incorporates the necessary values and ingredients to capture and grow market share and which utilize the principles of sustained or disruptive innovation including business model innovations. Herein we do not cover the "hot topic of innovation" *per se*, but we focus on the "secret sauce of innovation" which is the human capital and processes needed to create and deliver innovations to the market and capture value for the organization sustainably.

In addition to the list presented above we would add the following additional cultural traits of successful organizations:

- A focus on the market need first, which comes from being close to the customer or user
- Implementation of a reward system that values contribution and success and incorporates both psychological ownership of the outcome, and equity ownership
- Embraces an open innovation model to take advantage of ideas and collaborations beyond the "borders" of the company itself

The second challenge is to *imbue in this culture the following values* as identified in a recent Harvard Business Review article by Steven Prokesh, entitled "How GE Teaches Teams to Lead" [2].

- Challenge and involvement
- Freedom
- Trust and openness
- Time for ideas
- Playfulness and humor
- Conflict (creative tension but not destructive)
- Idea support
- Debate
- Risk taking

These common principles form the basis for building a managerial team and creative culture needed to innovate. A paraphrase from Phil Jackson, the most winning basketball coach in history is appropriate here; the strength of the team is each individual member — the strength of each member is the team.

KEY QUESTIONS TO ASK WHEN BUILDING THE TEAM — THE ACADEMIC PERSPECTIVE

Building a team is comprised of three phases summarized by Thompson[3], each of which must be re-visited as the organization and the team transitions from startup to development and commercialization stage, and then proceeds to market launch, growth and maturity.

Phase One consists of **Task analysis**. Specifically what is the work that needs to be performed and what is its focus, how much authority and autonomy does the team have to manage its own work, what is the degree of interdependence among the team members, and are the team members interests aligned or competitive?

Phase Two consists of the **People required** to perform the tasks to achieve at least the next milestone or two. How many people are needed, what technical, task management, and interpersonal skills are required, and what diversity is optimal for the team?

Phase Three consists of **Processes and Procedures required** to achieve success. What are the explicit or spoken norms, what are the implicit norms, which norms are conducive for performance, how are ineffective norms revised, and how much structure is required?

Overlaying these tasks, people and processes is the entrepreneurial culture that is desired. That is, those organizational characteristics and norms noted above plus the "expected entrepreneurial style of the people engaged," e.g. willingness to assume "some" risk, thriving on chaos, not controlling, positive, passionate, perseverant, and motivated to make an impact, or even to change the world.

BUILDING A BIOTECHNOLOGY ORGANIZATION

Most early stage biotechnology companies, as with most technology companies, start with two or three founders. They bring their passion, vision and mantra for a new company, along with the needed expertise, skill sets and networks to provide leadership for the two key and critical dimensions, each with its own attendant risks: 1) *technology advancement, and 2) business/market development*. In effect, upon founding, the *task analysis* and *people required* phases occur simultaneously and the founders form the kernel of a viable startup. Following the Thompson framework, the focus on developing and advancing the technology and the market in parallel is the essential "task" to be done, and the founders are the "key people who perform those tasks" — these are organization specific. This initial founding team (and their advisors added as necessary) then evolves through

Phases One and Two of the Thompson "model" in parallel where team members are acquired to evolve the technology and the market/industry dimensions, while advancing the commercialization process and developing the business model. It is understood that they must also acquire the needed financial resources to move forward. In most biotechnology startups where both technology leadership and business leadership are essential, decisions are most often made informally and by consensus, with input and perspective from both dimensions, but over time these roles evolve into a more formal structure, with decisions by the Chief Executive Officer (CEO) and Chief Technical or Scientific Officer (CTO or CSO). In most technology-enabled organizations (including biotechnology) the task analysis indicates that leadership is required to:

- Provide vision, strategic direction, fund raising, team building and overall leadership
- Lead scientific advancement, technology commercialization, and product development
- Lead business development and partnering

Additionally for biotechnology companies specifically, it will be necessary to add the capacity to deal with the following activities:

- Regulatory compliance and clinical demonstration
- Intellectual Property (IP) development
- Reimbursement

Early on these tasks can be accomplished by the members of the founding team and/or by part-time talent. These people expand from the kernel to comprise the core of the startup and development-stage team that most often has several people with perhaps two C-level positions designated to handle both inside and outside functions — these include simultaneous development of the product while working in parallel to more thoroughly understand and address customer/user need and the external environment. Acquiring people assets in biotechnology/tech companies is as important as acquiring financial assets, but one is required to accomplish the other — while advancing the opportunity and proposed solution. In effect an additional key task is developing the organization — most often consuming a significant part of the CEO's time allocation, along with acquiring funding! A quote here is appropriate to consider when building and growing biotechnology companies which are knowledge-based organizations. "Your most precious possession is not your financial assets. Your most precious possession is the people you have working there, and what they carry around in their heads, and their ability to work together" — attributed to Robert Reich, former Secretary of Labor in the Clinton administration and now a professor at the University of California, Berkeley.

The team that comprises an early-stage organization is not complete without developing its "periphery" — team members who serve in a more advisory function and contribute to the organization on an as-needed basis. It is critical for early stage organizations to develop a set of directors/advisors that bring specialized expertise, connections, and access to networks for funding, partnering, hiring, etc. Note that an Advisory Board and a Board of Directors each perform different functions. Most important is the need to institute a formalized, but small Board of Directors (BOD) of at least three people, including independent director(s) perhaps growing to five as equity investment occurs. The BOD provides overview of strategic direction and operations but also ensures that corrective actions based on internal and external changes and issues are addressed in a timely manner. The BOD will have fiduciary responsibility and other boards are advisory only — most often providing specialized knowledge and guidance such as science/technology, clinical development, etc. These bridges between the internal organization and the external environment also provide credibility and validation of the opportunity being pursued via the reputation of the people engaged with the organization. These directors and advisors are most often compensated via equity using industry norms as guidelines for a directors and advisors stock option pool.

Therefore the leadership team consists of core members (i.e., "the people on the ground) who have committed and are willing to take a risk to join the company, and the peripheral members, the BOD/Advisory Board. This extended team is expected to provide expertise, networks, perspective, and discipline as follows:

- Access to people, capital, partners, and markets/customers
- Access to counsel and expertise for IP, regulatory, reimbursement, clinical trials, corporate agreements
- Advice, experienced perspective and mentoring
- Adherence to plan and fiduciary responsibility

As noted above the characteristics of this extended team include the knowledge, skills and expertise, coupled with the requisite interpersonal skills (diversity, collaborative and communicative), and who have a shared value system (a common purpose and vision, trust, and sense of humor). In addition since in many technology-based organizations one is dealing with large egos, it is advisable to be able to "check your egos at the door."

FINDING AND HIRING GOOD PEOPLE

Finding and hiring good team members is the most important challenge faced by any company let alone a start-up or early stage organization! Especially at the earliest stages of any organization the CEO and other founders must be personally engaged in the hiring process since the "organizational DNA" or culture is imprinted starting with the hiring process. Selecting the right people "with the right DNA" is important to building the desired cultural norms — both spoken and unspoken. All startups should strive to hire only "A players" since excellence is essential to company success. Don't just hire to get the job done, make sure that the person "fits" and can also do the current task or job as well as grow with the organization. Hiring is expensive and time consuming, so hire right. A bad fit can be bad for the organization and replacing someone is also problematic and expensive. But if replacement is necessary do it quickly and professionally otherwise the "bad fit" will affect the organization itself.

Diversity is good since there are many skill sets required to build a successful company and diverse perspectives and experience sets provide more enlightened and innovative solutions. In a biotechnology or biomedical company diversity includes: various scientific backgrounds, business development/industry knowledge, expertise ranging from IP to regulatory to reimbursement. Additionally one must deal with perspectives gathered in small companies and in larger, more mature organizations, e.g. pharma or large medical devices companies. All of these key elements of the extended management team need to be integrated into the entrepreneurial and innovative culture being built. We advise embracing diversity, but not relying on chance to develop synergies as the team is built up over the life cycle of the company. It is also important to build mechanisms and processes to manage diversity not only internally, but also across the boundaries of the firm as networked innovation and partnering emerge as a norm in the biotechnology/biopharma industry. This open innovation business model is becoming increasingly important as industry convergence continues, blurring the boundaries of what is pharma and what is biotechnology. The team, culture and vision sharing are as important as skill sets so that there is trust, liking, and respect (unspoken norms) across the team and organization. Most successful organizations build this mentality into the hiring process and walk away from talented people if the cultural fit is not there.

It is important to understand and deal with factors that motivate entrepreneurs, and to address them individually as the team is built and expanded. It is important to know what motivates each member of the team. Entrepreneurial characteristics that are pertinent to biotechnology companies have been discussed by Boni in his review of Binder's book[1].

WHAT MAKES TEAMS WORK, OR NOT? THE ACADEMIC PERSPECTIVE

To discuss what works and what does not, we need to deal with three key factors. 1) the structure of the team, which includes roles and routines; 2) behavioral integration — managing the diversity; and 3) team norms — goals and shared values, and means of coordinating, communicating, managing conflict, making decisions, running meetings, and norm enforcement.

Larson and LaFasto[4] list the following necessary conditions for effective teamwork.

- A clear, shared and elevating goal
- A results-driven structure, that includes
  - Clear roles and accountabilities
  - An effective communication system
  - Monitoring of individual performance and providing feedback
  - Fact-based judgments
- Competent team members (technical and interpersonal)
- Unified commitment
- Collaborative climate
- Standards of excellence
- External support and recognition
- Principled leadership

We refer the reader who is further interested in building effective teams to several good *Harvard Business Review* articles by Billington[5] and Katzenbach and Smith[6]. While these articles are not targeted specifically at knowledge-based biotechnology companies, the authors ask the question of what makes the difference between teams that perform and those that don't? These are universal lessons. In that regard they point out that teams and groups are not the same. The team is defined as "as small number of people with complementary skills (competence) who are committed to a common purpose, set of performance goals, and an approach for which they hold themselves mutually accountable." The Billington article points out that mutual accountability differentiates a team from a group. In a team if the team fails (or the company), all fail together. If the team succeeds all are rewarded. One other best practice that is important in any startup organization is for the leadership team (and its Board) to establish and maintain a sense of urgency. Kotter identifies the sense of urgency as the first and essential step in his 8-step process for leading change identified in extensive case studies [7]. From a practical perspective we advise that the team consider spending a lot of time together outside of the workplace and inside (which is inevitable in a startup environment).

SUMMARY OF "LESSONS LEARNED" FROM THE BOOT CAMP PANELS TARGETED AT BIOTECHNOLOGY COMPANIES

It would be remiss on our part to leave the reader with the above summary without discussing how the pedagogical approach taken at the Boot Camp and what has been learned over the years from multiple panels. After an introductory discussion of the above principles we assemble a panel that consists of founders, key officers and investors in an emerging biotechnology company. These panels are different at each Boot Camp, so the following represents a summary of the "hot topics" that are consistently discussed at these panels.

Each panel is charged by the moderator with discussing a number of key issues and exchanging views on how they are handled "in the real world" during the startup and development stages of the organization. Most organizations represented are still at the development/clinical level or just entering the market growth stage. The audience has an opportunity to ask questions and to engage in discussion with each other and with the panelists and Moderator. Over the years the following topics and brief summary of recommendations gather the most questions and discussion:

1. Virtual Startups vs. "Bricks and Mortar"

Acquiring capital is very difficult if not impossible until a considerable amount of risk (technology, IP, clinical, and team) is reduced. However, progress must be made to interest investors. Therefore, founders most often need to acquire non-equity resources (government, economic development) or funding from individual angels to raise limited seed capital. It is also beneficial to do so to increase market cap and reduce dilution. So reducing the amount of capital needed is recommended by leveraging resources; e.g. use of academic facilities, outsourcing product development and clinical work to others. It is recommended not to invest in facilities except for the bare minimum, instead invest in key people who may or may not join the company full time, and have founders fill multiple functions on the management team. "Cash is king" so use it wisely. Eventually you will need some facilities so consider locating in an incubator or leveraging common space with existing organizations in a research park. Invest in "hard assets" only when this is justifiable after evolving down the commercialization path.

2. The First Hires and Building Boards

Start with a small core team that originally consists of the founders (2 or 3) and a few part time consultants — noted above. Identify key advisors and directors who can provide you with good advice and credibility and pay them with equity (it's worth the dilution). Find an attorney that will work on a contingent basis (not always possible) and consult with them on creative and legal ways to handle compensation, stock and corporate partnering issues. However, make sure that vesting is used for stock options. Alternately, issue restricted stock. Consider what happens to the stock if key people leave? If it is gone with the departing person it will dilute those who remain since the person has to be replaced. Pay the core team less-than-competitive salaries until funds are raised — they will make up for it with stock. Hire for the essential tasks that need to get done but make sure the fit is good (see below). Use mentors to help you and to locate advisory board members and directors. Initially you should have no more that three to five science/business advisory board members (non fiduciary positions) and three board members (with at least one outside, credible and experienced person). Advisors and directors surrounding and supporting (and mentoring) the core team will facilitate progress with commercialization and will lead to downstream success with fund raising and partnering.

Hiring progression/priority will generally proceed in the following order of priority:

- Business, scientific/technical, market/business development leadership team
- Clinical/medical, regulatory and IP expertise (can be outsourced with inside leadership via a key employee at the appropriate time)
- Personnel to contribute to the scientific and business agenda associated with commercialization according to the organizational priorities (generally product development and customer/user development).
- Financial management (once significant funds are raised, especially A-round financing)

Keep in mind that as new team members join they must buy into the culture that has been created by the founders. So these first members are key and will build and preserve the corporate culture. Fit, shared values, relevant experience and ability to execute are all important.

A subset of this discussion always revolves around the issue of "splitting the equity pie" and dilution. One could write an entire article on this topic; suffice it to say that the initial equity should be split among the founders and early hires (if any) based on what they have contributed to the company formation, what they will contribute going forward, and the level of risk each person takes. Engage a good lawyer to help with this because it is always a contentious issue as to who contributed what and who will do so going forward. Most prominent among the contentious issue is the debate about the weighting of science vs. business in equity participation. Make sure that the founders get rewarded for their founding contributions (value is attributed to both technology and business acumen), and make sure that those who take the risk and actually join the company are rewarded for that as well. Small equity pools are then created for advisors/directors and for stock options for employees to be hired prior to the next funding tranche — typically 15% to 20% of the total for all parties. The pool will then be replenished prior to the next equity raise (investors will most often insist that the dilution will be taken by the insiders and not the investors). Many entrepreneurs worry excessively about dilution. For those who have been through this many times however, there is a realization that creating value that builds the capitalization of the company is the key outcome to be pursued, and taking outside money is essential to value creation and risk reduction, i.e. a "small piece of a large pie" is better than the alternative. Getting to the end game is the objective!

3. Balancing science and commercialization

In biotechnology and biomedical companies there is always the need to continually advance the science (to prove principle, build the platform and the IP portfolio). However, progress down the commercialization pathway is necessary to generate the funding that will be needed to attract subsequent team members. Therefore, priorities and a sense of urgency to advance the technology and business have to be established early on at the founder and board level and managed carefully by the CEO and CTO or CSO of the company. Many organizations maintain close ties to a university where scientific advances can be handled (but be careful of IP and conflict of interest issues). Commercialization involves clinical demonstration in parallel with product development, which is difficult in a regulated environment. Once funding is raised make sure to allocate a small portion to advance the science and also consider some government augmentation (via the SBIR program) to achieve those objectives. While SBIR funding is non-dilutive, sometimes the timing is not consistent with commercialization priorities.

4. Managing through transitions

Along the commercialization pathway company leadership and the board will need to deal with evolution of the team as people join the team and leave the team — either voluntarily or involuntarily. Sometimes founders take "lesser roles" as new leadership is required to move forward thru the clinic and into the marketplace, or to raise venture capital and/or partnership funding. We have not found the perfect formula for dealing with these issues. One thing that can be counted on is that it will happen in virtually every company. In order to manage this process the right people must be on the board or on the advisory group to assist with the people issues — the addition or subtraction as well as the team remaining. Nothing can destroy team chemistry faster than a mismanaged transition. The best advice is to handle the situation quickly and professionally with good communication to all of the constituencies of the company appropriate to the specific situation. It is rare that a founder who becomes the CEO of a biotechnology startup can survive through to the acquisition or IPO.

In conclusion, we present one current and timely thought that is becoming increasingly important for building and funding biotechnology companies. As these organizations are built, consider using capital efficient business models. This is essential for biotechnology and biomedical companies where it is important to reduce technical, market, and team risks prior to bringing in the extensive amounts of capital required and even to form win-win partnerships while maintaining the ability to share significantly in the value that has been created by the team. Iterative product and market development can lead to lower capital expenditures and faster time to market (even though the regulatory authorities tends to slow down the cycle time — this is not as much of an issue for other technology companies where lean and agile methods are being employed). There is much discussion in the field of biotechnology about the use of leveraged capabilities and assets including the building of "virtual companies" using management teams that have prior experience with bringing products to market, and/or by partnering with outside organizations. Also consider creating value and reducing risk via proof of principle demonstration in a clinical setting (even if off shore) prior to raising large amounts of capital. An extensive discussion on this topic is beyond the scope of the current article, however suffice it to say that virtual companies can be created to leverage expertise by using open innovation principles to partner for technology, market access, product development and clinical testing, manufacturing, even management teams. Why build capacity that already exists? Sharing value might be a better option. The challenge is to build a core team that is equipped with the processes and networks to access and effectively manage these relationships. However, keep in mind that it will be necessary to have expertise on the extended team to manage the partnered or outsourced tasks. This will require the existence of talent that has experience with product development, clinical testing, etc. The subject of building teams in open innovation environments is a topic of current work and research.

Keep in mind that the overall objective in building a team is to address one key component of risk reduction for the organization — demonstration of the ability to execute. The team addresses market risk, regulatory risk, IP risk, and risk associated with reimbursement. Other sections of the Boot Camp deal with reduction of technical risk and are not addressed explicitly herein except how they are addressed by having the right people on the team at the right time.

REFERENCES

1. Boni, Arthur A. (2009), "Science Lessons" What biotech taught me about management, A Book Review of Gordon Binder and Philip Bashe. J. Commercial Biotechnology, Vol. 15, 1, 86-91. Also see Binder and Bashe, Harvard Business Press (2008)
2. Prokesh, Steven (2009), "How GE Teaches Teams to Lead Change", Harvard Business Review Reprint R0901J.
3. Thompson, L. (2011), "Making the Team. A guide for managers." 4th Edition, Prentice Hall
4. Larson, Carl E. and Frank M. J. LaFasto (1989), "Teamwork: What must go right, what can go wrong", Sage, Newberry Park, CA.
5. Billington, Jim (1997), "The Three Essentials of an Effective Team", Harvard Business Review Reprint U9701A
6. Katzenbach, Jon R. and Douglas K. Smith (2005), "The Discipline of Teams", Harvard Business Review Reprint R0507P
7. Kotter, John P. (1996), "Leading Change", Harvard Business Press, Boston, MA

The pitch and business plan for investors and partners

Received: November 8 2011

Arthur A. Boni
is John R. Thorne Chair of Entrepreneurship; Distinguished Career Professor; and Director, Donald H. Jones Center for Entrepreneurship, Tepper School of Business, Carnegie Mellon University.

ABSTRACT
This article covers the essentials of constructing and delivering a "pitch" of a business opportunity to potential investors or corporate partners. We advocate constructing an effective pitch first and then using that as a guide to prepare your business plan. The content of the pitch itself as described herein in effect comprises the elements to be incorporated into a business plan as a more comprehensive documentation of the business. Therefore, focusing efforts on creating and delivering the components of a compelling pitch will provide the essence of what you will then need to put into prose as a business plan. So, our mantra is "pitch then plan," i. e. don't waste your time writing prose until you know what you need to write. Furthermore plans change many times between the first writing and successful implementation.

Journal of Commercial Biotechnology (2012) 18, 38–42. doi: 10.5912/jcb.509
Keywords: pitch; business planl investor; partner

INTRODUCTION

PITCHES DO NOT come easily, so our recommendation is to work up a pitch and practice it multiple times with multiple audiences. This is an iterative process and as you proceed to try out your pitch with diverse audiences (customers, partners, investors, team members), you will receive questions and constructive feedback that will in effect help you to refine your understanding of the market need, how to articulate the value that you create, your go to market strategy, etc. Plans don't sell opportunities; people do via their pitch. Investors typically invest in management teams — especially those with good opportunities and an ability to articulate that opportunity. It is well known by most experienced investors that a business strategy changes many times from inception to successful execution. Therefore, at the "end of the day" most will invest in a team that can survive, grow, and thrive in the face of incredible uncertainty and risk. The team is so important that our Biotechnology Entrepreneurship Boot Camp dedicates a special session to this topic (see article by Boni and Weingart in this special edition).

In the Biotechnology Entrepreneurship Boot Camp we assemble two sessions to demonstrate how to give a pitch. The first is targeted at venture/angel investors and the second to corporate partners. We assemble panels to which CEOs give their pitch, and then get "grilled" in real time by real investors and corporate partners. Seeing this process in action provides a great learning experience for both the CEO and the participants in the boot camp. This article will illustrate that the pitches basically contain the same elements, but that the elements of focus for a venture pitch and a corporate pitch may differ somewhat. Distinctions will be made in the article. Unfortunately we cannot simulate the real time learning experience in this article, but the information contained in this article is provided to the audience, to the presenting CEOs, and to the VC/partner panels prior to the Boot Camp itself.

The setting for the pitch is that you have been able to get an introduction to an angel investor, venture capitalist, or potential corporate partner. This personal introduction or referral is generally required, especially for a venture or angel investors since they are deluged with plans and requests for audiences. They are not going to be able to spend the time reading through a business plan (or perhaps even an executive summary) to decide whether or not to meet with you. It is also helpful with

Correspondence: Arthur A. Boni, Ph. D., John R. Thorne Professor of Entrepreneurship, Tepper School of Business, Carnegie Mellon University, Pittsburgh, PA 15213 US. Email: boni@andrew.cmu.edu

the corporate partner since it may not be clear whom to contact in a large organization. Therefore, an introduction from someone whom they know, trust and respect will help with setting up a meeting and also prevent your hard work from ending up "in the wastebasket or shredder." There will generally be a short phone call to the appropriate person in the organization and this is your first opportunity to deliver your "elevator pitch." The ability to deliver a short compelling summary of your opportunity sets the stage for engaging your audience. We cannot stress enough the importance of being able to do this in the short amount of time (comparable to a ride on an elevator) when you encounter someone for the first time. Your passion and dedication need to come through. It's not the pitch itself, but your ability to articulate your venture concept, why you think it is different and will succeed, and what you plan to do in a few short sentences. You can convey the importance of the problem, the uniqueness of your solutions, and the critical skills that you and your team will bring to the success of your opportunity — and why this would set the stage for making money and for the corporate partner to find a significant opportunity for their portfolio. The purpose of the elevator pitch is to convince them that they want to hear more in a meeting and perhaps request some preliminary information such as an executive summary prior to the meeting.

Now, at the meeting, you must convince them that your opportunity is a good one for them. You won't get a second chance to make a good impression. Therefore, before the meeting you should prepare yourself as follows:

- Know your audience and *prepare for high-stakes selling.* Anticipate the questions that may be asked and prepare answers to them. Do some market research on the firm and see what deals they have done, what they like, what they don't like. Talk to someone who has done business with the firm previously including the contact that you used to get the meeting.
- If you are talking to a venture capitalist be prepared to *pitch a business and not a product.* They invest in businesses. Corporate partners tend to be more interested in products since most typically the partner will bring the product to market thru their organization. However, they will need to be convinced that your business is a viable partner that will be able to deliver the product to them over time.
- *Stay short and focused.* Guy Kawasaki[1] in "The Art of the Start" suggests that your pitch should consist of 10 slides, 20 minutes, and 30-point font. You can generally make all of your points on 10 slides, but you may need additional slides as backups to get into detail as questions are asked. A first meeting will generally not last more than an hour, so 20 minutes of planned presentation will expand to an hour or so with questions and discussion. A 30 point font means that you should not cover a slide with so much material that it really can't be read by someone with normal vision. Use pictures or simply graphs/charts so that they will listen to what you are saying (instead of reading ahead and getting confused). You should know the key points that you want to make so that you don't need to write down every point that you will want to make verbally.
- Kawasaki[1] also advises that the entrepreneur ask himself or herself the question *"so what?"* That advice is particularly pertinent to the technologist who assumes that the audience is as familiar with their jargon as they are. Anticipate that the audience may or may not understand the significance of something that you understand very well but that they do not. So anticipate this and explain the significance of what you have just said in terms that they understand with a "for instance" to really make your point.

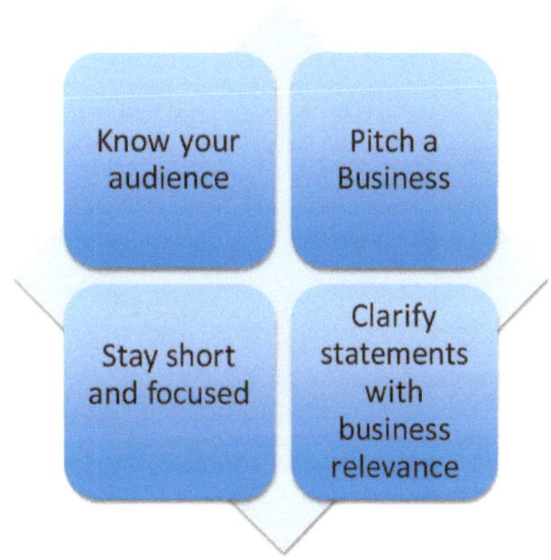

Figure 1: Preparing for the pitch

THE ANATOMY OF A PITCH

At the start of your pitch, the first objective will be to get the attention of the audience and have them in effect anticipating to hear more about the opportunity and your team. This is the objective of the Elevator Pitch. This first part of the Elevator Pitch should be short and comprise the first "15 seconds or so." This first part is to engage the audience and make sure that you highlight what you do, explain why it's important, and say something that validates the authenticity of the opportunity. Hopefully the audience will now be engaged and want to hear more. You then proceed to the next part of the elevator pitch, which should take about 60 seconds more. It should cover the following points:

Figure 2: The ten necessary slides

- The opportunity is big and unsolved and the need is compelling
- Your solution is unique (can be differentiated) and has a competitive advantage
- Identify the customer and why they care — talk about the value that you create (or the pain that you take away)
- How you are going to make money (and return it to the investor)
- Why this CEO and team?

In the elevator pitch and in the pitch itself (which follows) you will need to be enthusiastic, passionate, and authentic. If you don't have an answer to a question don't fake it. Just say that you'll get back with them, and then do.

Your pitch should be organized into ten sections as shown below (remember Kawasaki's 10 slides). These topics should look familiar to you since they comprise the main elements of any business plan as outlined in any book that includes a discussion on construction of business plans; c. f. Timmons and Spinelli [2], Kaplan and Warren[3], Dorf and Byers[4] for example. The slides (as do you) need to communicate enough information to explain your material and also enlist intelligent questions from the audience, while encouraging them to the next step in evaluating your opportunity in subsequent meetings and in the due diligence process. Keep in mind that you will not walk out of this meeting with a deal in hand. Moving forward to a term sheet (and deal) is an iterative process and will take a period of time. VCs and corporate partners each have their own process that takes months, not days.

1. **Title Slide**. This slide lists the name of the organization, address, contact person and any other pertinent information that your audience will need to know who you are and how to contact you. This slide is just used as a visual while you give your elevator pitch. Some people may include a picture or some other object on this slide that illustrates something about them or their opportunity.
2. **The Problem**. This slide should be used to illustrate a clear and compelling market need. What is the "pain" or compelling need in the marketplace? How is that need addressed today, what gaps exist, and who are the customers? The goal here is to get everyone in the room to buy into the problem or need, that it really exists, and that it is significant. Often time some presenters use a "persona" to illustrate the problem in the form of a fictitious or real person who can be used to make the problem personal in the context of people who we really know. Make it real!
3. **The Solution**. Simply said this slide explains how your proposed solution solves the problem, how it works, why is it different, and what are the demonstrated outcomes of the concept as it currently stands. Why is this a better solution than the alternative? For a partner, get them thinking about synergies with what they do and what they need. For an investor, get them thinking of how this solution will result in fast customer acquisition and market penetration.
4. **The Business Model**. The business model can be defined as how you arrange the elements of your business to create, deliver and capture value in the market. There are four elements: customers, your offering (product/service), the infrastructure, and the financial viability. See Osterwalder[5] for a very compelling and visual description of how business models are

constructed. You will need here to talk about who is your customer, how many are there, the value that you create, how it is different, significant and sustainable, your value proposition, how you access customers (access to market channels), and how you make money (the revenue model). For the investor this slide will get them thinking about exit paths through M&A or IPO. For a corporate partner it will show where you sit in the value chain and where they sit and capture or share value.

5. **The Management Team**. Most entrepreneurs want to address the management team early in the presentation since the management team is so essential to the opportunity. You will want to address the issue of "why this team" in this section. Identify who is on your team, list their relevant experience(s) and expertise that is needed to advance your opportunity. Also list your Board of Directors and Advisors and note your existing investors if you have any. We recommend that you recognize gaps in your team and address how they will be filled — it is better to anticipate this question proactively.
6. **The Technology**. Explain your technology in simple terms, how it is unique and protected (yes patents are essential in biomedical companies in particular and also in most technology companies), current status of your IP, and the current state of development and demonstration in the lab, clinic, etc. Are there other technologies in development that might be competitive or synergistic? The less text here the better. Use charts, pictures, etc. Have backup charts or white papers available as backups.
7. **The Market**. What are the social, economic and technological factors (SET Factors) or drivers that create this market opportunity? Who are your target customers? Talk about market size, growth rates, market segments, market entry points, what will drive adoption, and how you acquire customers. In this latter regard how do you drive customer awareness, consideration, choice and retention? How much does it cost and how long will it take to make a sale? You can also talk about regulatory and reimbursement issues in this section.
8. **The Competition**. Overview the competitive landscape — who is in this market, what are the product lines, competing technologies in development? Show how you compare / differentiate against the competition, state your barriers to entry and reiterate your sustainable competitive advantage. Also anticipate how your competition will respond to your market entry. Keep in mind that there are always competitors even if it's getting the job done in a different way.
9. **The Finances**. You will be expected to produce a five-year pro forma with profit & loss, cash flow, and balance sheet (for the business plan). Have these in a back-up slide, but in this section use a chart to show revenue and cash needs over time. Some also use this chart to illustrate funding needs, tranches, and value inflection points. You will be expected to address key metrics such as customers, products sold, market penetration rates, etc. appropriate for your business. You might also want to have a backup chart to list key assumptions that you have made.
10. **The Current Status**. Explain what you have accomplished to date, and what still needs to be accomplished (tasks and current milestones) — development, IP, clinical, team, etc. Indicate how much money you will need over time and how you will reduce risk in technology, team, market, clinical, etc. Talk about how risk is reduced and value created at each milestone over the 5-year period (investors think in terms of value created for each funding tranche). Talk a little about how you are building value in your company, but be aware that investors will be thinking about how to monetize that value in an exit — what IRRs are possible here? You may also want to start a discussion here about next steps with the investor or corporate partner.

As final advice we recommend that you practice your pitch repeatedly, learn to pitch at the right level to get the next meeting. Sell the team, the opportunity and the synergies for the corporate partner. Have lots of backup slides for answers to the questions (and have them organized and indexed for easy retrieval)! Then use these to write your business plan. But don't put it into a bound volume since it will change many times over the lifetime of your venture. Things change frequently and unexpectedly in the real world.

THE BUSINESS PLAN

The above can be assembled into a business plan that demonstrates that your business is based on the following principle anchors as described by Timmons and Spinelli[2]:

1. Creates or adds significant value to a customer or user
2. Solves a significant problem in a large and growing market
3. The opportunity can be differentiated and a sustained competitive advantage can be developed
4. The market has the potential for good margins and moneymaking characteristics
5. There is a good fit with the founders and management team at the time with a balance of risk and reward — and alignment of interest with all constituents including investors

The plan itself may be visualized in three layers:

- The Foundation or Pillars of the Business (Opportunity)
    - Opportunity: The Problem (need), The Solution (your offering), The Business Model

- The Infrastructure that enables execution (Resources)
    - Resources: The Management Team, The Technology, The Finances, Current Status

- The Context (Social, Economic and Technological Drivers)
    - The Market, The Competition

The descriptions of the appropriate content for the business plan are outlined in the above section on the pitch. These sections may be augmented with Appendices and supporting documentation drawn from the backup slides used in your slide deck for the pitch.

In concluding it is important to stress that biotechnology investment opportunities fall into the general class of entrepreneurial deals. Deals are done every day in all sectors of the economy and there is extensive literature on what constitutes a good deal and what does not. That is beyond the scope of this article. However what is important to stress here is that in biotech deals science and technology is very important (and must be differentiated carefully), however, the quality of the deal is based on many more factors, and these must be evident in your pitch and your plan. Above we have stated the importance of the team (and its advisors). A good financial deal is good for all parties concerned so interests must align well — at the time of the deal and into the future. Financial return and its magnitude and timing are important for investors, winning products are important to partners, and equity ownership is important for entrepreneurs. A good deal is further differentiated by many factors. These include with respect to the founders and management team, but are not limited to: ability to communicate, passion, conviction, perseverance, willingness to be coached, and ability to work in a collaborative team environment that incorporates diverse skills sets and perspectives.

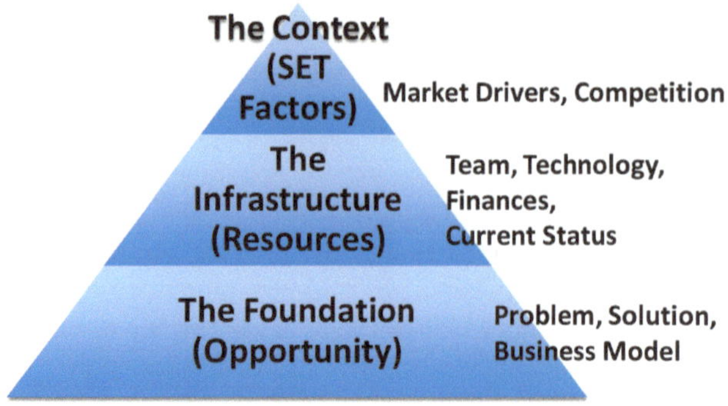

Figure 3: Building blocks for a winning business

REFERENCES

1. Kawasaki, Guy, "The Art of the Start", Portfolio, published by the Penguin Group (2004)
2. Timmons, Jeffrey A. and Spinelli, Steven, "New Venture Creation: Entrepreneurship for the 21st Century", McGraw-Hill Irwin (2007)
3. Kaplan, Jack M. and Warren, Anthony C., "Patterns of Entrepreneurial Management", John Wiley and Sons, Inc. (2010)
4. Dorf, Richard C. and Byers, Thomas H., "Technology Ventures, From Idea to Venture", McGraw Hill (2008)
5. Osterwalder, Alexander and Pigneur, Yves, "Business Model Generation", John Wiley & Sons, Inc. (2010)

Strategic engagement of the science-business media

Received: November 14 2011; revised February 14 2012

Moira Gunn

founded and directs the Business of Biotechnology Program in the School of Management at the University of San Francisco, where she is an assistant professor. A former NASA scientist and engineer, she is producer and host of Tech Nation and its regular segment BioTech Nation, which air on the National Public Radio 24-hour Program Stream, among other venues. Dr. Gunn holds a patent in human nutrition research, and was awarded the 2011 National Science Board's Public Service Award to the Individual for her contributions to the public understanding of science and engineering.

ABSTRACT

As surely as the bio-enterprise can benefit from positive media coverage, it cannot thrive in the face of unanswered negative and/or inaccurate media attention. On all counts, the bio-enterprise must be able to strategically engage with the media at every stage in its life cycle. This article describes the global science-business media landscape, including traditional media and emergent social media and information in the online space. Current research is used to document the interdependence of media, how sources of information feed media coverage, the challenges of science communication in the broader context of business, and the effect of media engagement by the CEO.

A strategic model is presented which relates the bio-enterprise to global media, providing a larger framework within which to develop media action plans at critical junctures in the life of the bio-enterprise. Further documented is the difference between journalistic and non-journalistic media, and how journalistic ethics and standards guidelines work with ethical persuasion practices to the benefit of the bio-enterprise.

Journal of Commercial Biotechnology (2012) 18, 43–54. doi: 10.5912/jcb.518

Keywords: bio-entrepreneurship; science-business media; science-business journalists; biotechnology media; media strategies;media model

INTRODUCTION

As surely as the bio-enterprise can benefit from competent, positive media coverage, it cannot thrive in the face of unanswered negative and/or inaccurate media attention. Less well understood is how the bio-enterprise relates to global media, what the bio-enterprise can and cannot control, and how bio-enterprise management qualifies specific tactics, activities and postures recommended by media relations professionals, both internal and external to the organization.

This article describes the global media landscape, offering insights into the sources of information which feed media coverage, the difference between journalistic and non-journalistic media, the increased challenge of biotechnology industry reporting, and how ethics and standards guidelines work with ethical persuasion practices to the benefit of the bio-enterprise. A strategic model is presented which relates the bio-enterprise to global media, providing a larger framework within which to develop media action plans specific to target audiences at critical junctures in the life of the bio-enterprise.

Correspondence: Moira Gunn, University of San Francisco School of Management, San Francisco, CA, US. E-mail: gunn@usfca.edu

Biotechnology and the Global Media Landscape

Science-business media coverage began in earnest with the October, 1980 coverage of the initial public offering (IPO) of Genentech by the traditional media, followed closely by the March, 1981 IPO of Cetus Corporation. Since that time, science-business media coverage in the biotechnology sphere has been largely the province of financial coverage of publicly-traded biotechnology stocks, peer-reviewed research which has potential for business applications, popular science magazines, and the occasional feature in newspaper science sections. Still, governmental policy decisions, strong public reaction to biotechnology, and "color" stories, such as "Dolly the sheep" do periodically capture the attention of mainstream media. Today, a vastly-expanded complement of information spaces exists, the direct result of the ubiquity of the Internet and readily-available online and social media tools.

Professional media coverage continues to be provided by professional science-business journalists through traditional media outlets. These traditional outlets have also developed an online presence with some solely existing in the online space today. Yet, this is only a small part of the total picture — at least, by volume of information.

The greatest growth in information is online and originates from the bio-enterprises themselves, along with relevant government agencies, research institutions, healthcare providers, industry organizations, NGO's, advocacy movements, and other relevant organizations. Still, other online information is created by individuals. These contributions come from a remarkable base, ranging from the perspectives of credentialed scientists and industry experts, to volunteer coverage by interested professional science-business journalists, to the un-sourced and unedited conjecture of anyone who cares to post in the online space.

Moreover, in the online space, any of this information can be aggregated, re-referenced and re-framed via commonly-available Internet tools, all outside the control of the originator and/or the bio-enterprise. All of this information is accessible to anyone with Internet access; much is free; none regulated.

Interdependent Media

Traditional journalistic media outlets and non-traditional online media are no longer independent phenomena; they have become interdependent, but in different ways. Research has shown that traditional media, as exemplified by the *New York Times* and the *Washington Post*, started routinely reporting on blogs in the 2002 timeframe, and transitioned to citing blogs as sources in the 2005 timeframe.[1] More recent research shows that traditional media, when citing blogs, tend to cite them as sources of opinion, rather than sources of fact.[2]

Blogs and bloggers, on the other hand, have displayed two tendencies. [1] The first is to be heavily reliant on traditional media for source material — in fact, just under half of blog citations cite traditional media. Thus, the sources of information which feed traditional media, reach deeply into non-traditional online media outlets.

In counterpoint, the second tendency of blogs is to avoid citing any sources whatsoever. Perhaps more importantly at this juncture is how bloggers view themselves. The maturation of perceiving themselves as journalists grew from 17% in 2006 to 45% by 2008.[3]

For the bio-enterprise, this must be strongly considered. First of all, there is little or no editorial infrastructure apparent for contributions in the online space. In journalistic terms, this means that there is no internal second-party questioning of information prior to publication, no second-party fact-checking, no published rules on what makes or does not make for a valid source, no requirement to frame the information within a total context, etc.

To be sure, serious and responsible reporting and analysis can and does originate in the non-traditional online space; however, in these spaces, it is not always possible to determine where that is. Casual reading in the online space does not readily reveal the competencies or agenda of the creator, the validity of its information sources, the origination of the funds which enables the effort to generate the information, or the editorial guidelines under which the information is published.

Yet, for the bio-enterprise to ignore the non-traditional online space misses the Internet's greatest value: The ability to aggregate and distribute again and again, to amplify the value of positive coverage. It also misses the complementary phenomenon: Misinformation and third-party perceptions of negative circumstances can be amplified, as well.

Effective media strategies recognize and take advantage of these interdependencies and the amplification properties of the online space.

MEDIA CONSIDERATIONS FOR THE BIO-ENTERPRISE

All successful media is based on trust — trust between the media outlet and its readers, viewers and/or listeners. This refers not only to the trustworthiness of the information itself, but also to its trustworthiness over time, its consistency to deliver the expected information in a form the reader/viewer/listener can understand and utilize for the purposes each has in mind.

Every media outlet deliberately focuses on specific audiences, although the audience defines itself, and reader, viewer and/or listener of any media is the final arbiter of how information will be used and for what purpose. With respect to bio-enterprise media coverage, this presents special challenges.

The Challenge of Communicating Science and the Bio-Enterprise

Science is at the base of every biotechnology business proposition, and the successful communication of that science, its transition into a product, the intellectual property it encompasses, and the perceived risk of the

Table 1: Sample media outlet audiences

Industry Analysts
Venture Capitalists
Regulatory Personnel
Biotechnology Industry Organizations
Service Professionals
Financial Advisors
Policymakers
Legislators
Insurance Providers
Healthcare Providers
• Hospitals
• Healthcare Systems
• Individual Providers
Industry Organizations
Educators
Special Interest Groups
Consumer Advocates
Consumers

business endeavor are all familiar themes in traditional business media. This expands into the consumer and other media spaces with the introduction of the biotechnology products themselves — from new treatments and diagnostics in the healthcare field to genetically-modified products in the agricultural, energy and industrial fields. Table 1 contains a sample list of audiences that media is attempting to reach.

Explaining key insights to these audiences requires the effective communication of science; however, these audiences are self-defining, and even within audiences, science literacy varies widely. This makes effective communication a challenge, as much for the media outlets, as it is for the bio-enterprise.

Science communication in context

Much has been written about science communication; however, most has been directed at getting the basic science right, i.e., communicated clearly and accurately to audiences of various science competencies.[4-5] Some has even been focused specifically on the communication of the science of genetics.[6] Unfortunately, this approach presumes that getting the source science information right solves all problems — for all, including the bio-enterprise.

In the case where science is to be communicated by journalists, the bio-enterprise needs to address the intercultural communications "gap" between scientists and journalists. This is of prime importance to the bio-enterprise as its scientists are naturally called upon to explain its science. Described from an American perspective in the late 1990's,[7] more recent research has further refined this insight, and importantly so for the bio-enterprise.

An extensive survey of the factors affecting science communication by scientists and engineers was undertaken by the Royal Society in the UK and published in 2006.[8] This time, the research distinguished between communicating with general journalists and popular science journalists. In rough numbers, of the 1401 scientists surveyed, 1 out of 5 experienced difficulty in communicating their research results to the general media, while only 1 out of 20 experienced difficulty in communicating with popular science journalists. It appears that in science journalism, experience matters.

This strongly suggests that interactions between scientists and professional science-business journalists are likely to be far more successful than interactions with the general media, whether they be professional journalists or otherwise. But difficulty in communicating is only one issue for the scientist who interacts with the media.

In the same timeframe (2005-2006), a five-country survey (France, Germany, Japan, UK, USA) of peer-reviewed research scientists was specifically taken to examine their experience with professional media within the previous three years.[9] While no distinction was made between general media and professional science media, the 706 stem cell scientists (of the 1354 total scientists surveyed) expressed the greatest concerns about the possibility of negative publicity (74%), the unpredictability of journalists (84%) and a risk of incorrect quotation (94%).

Further research would be required to distinguish these reactions when dealing with general media vs. professional science media; however, the numbers are too great for this distinction to explain these concerns away. Arguably, the reactions of these scientists have less to do with science communication and far more to do with scientists being unclear about the profession of science-business journalism.

Science-business journalism

As science is at the core of every biotechnology business proposition, the journalists who cover this industry must be distinguished from general business journalists. These science-business journalists must also meet the specific science communication challenge, and place the science-business proposition within a larger business context. The content they create appears in recognized traditional media outlets, such as those listed in Table 2. While these media outlets may not be perceived exclusively as business media outlets, they serve the bio-enterprise by reaching relevant audiences.

Important in understanding the value of content published by traditional media outlets is that they subscribe to a published code of journalism ethics. Many of these codes can be found online at the Pew Research Center's Project for Excellence in Journalism.[10] While

Table 2: Sample professional science-business media outlets

Primary Media Channel *[++ Online Presences]*	Professional Media Outlet
Print	Barrons, Bloomberg Businessweek, BioWorld, Discover, The Economist, Financial Times of London, Forbes, Fortune, Genetic Engineering News, Technology Review, New York Times (Science Section, Finance Section, Op-Ed Section), SCRIP, Science News, Scientific American, USA Today, and Wall Street Journal, plus Relevant Local/Regional News Outlets (Example: San Francisco Business Times)
Television	BBC, CNBC, CNN, Discovery Channel, PBS
Radio	National Public Radio Science News coverage, NPR Talk: BioTech Nation, Talk of the Nation: "Science Friday"
Online-Only	Science|Business, Dow Jones MarketWatch

Special Note: Table 2 is not intended as an exhaustive list of Professional Science-Business Media Outlets presenting original journalistic content. In addition, each has a substantive on-line presence, which also follows journalistic guidelines. For the purposes of this discussion, the online materials are considered to be part of the original publication. As professional media frequently migrates and expands to multiple formats, a media outlet is listed under its most well-known media format. It should be noted that professional media outlets often change focus and coverage parameters, as well as media formats, over time; however, these outlets tend to remain true to their journalistic standards.

differing each from the other in minor detail, they all subscribe to the general tenets of professional journalism.[11] In common, they have the perspective, which the Committee of Concerned Journalists (CCJ) describes thusly: "The central purpose of journalism is to provide citizens with (the) accurate and reliable information they need to function in a free society."[12] CCJ's core principles appear in abbreviated form in Table 3, and illuminate the perspective of the professional journalist.

As science-business journalists are not simply communicating science, but rather science in a larger business context, the reactions of scientists, which include "negative publicity" and "unpredictability," become comprehensible, once the principles under which these journalists operate are understood. Science-business journalists are not a conduit for the information that the scientists or the bio-enterprise wish them to deliver to a target audience. They are an independent source for confirming facts and delivering relevant, informed opinion.

In fact, it is posited that the very existence of these codes of ethics require professional science-business journalists to get the science straight. As Messner, et al, point out: "Traditional media will only choose and include sources that they view as trustworthy, truthful and knowledgeable."[1] This becomes even more important for the bio-enterprise when considering the recent reductions-in-force in professional science and technology reporting staff in traditional media.[13] Reductions in qualified and capable science-business journalists places pressure on the collective of bio-enterprise as a whole to influence the global media landscape.

The emergence of social media

Professional media outlets have always adopted new technologies, as exemplified by radio, television and satellite communications. These technologies have been sufficiently expensive to deliver that few could afford to generate and disseminate information. Today, via the Internet and freely available World Wide Web tools, any-

Table 3: Core principles of journalism: Committee of Concerned Journalists[13]

Journalism's first obligation is to the truth.
Its first loyalty is to citizens.
Its essence is a discipline of verification.
Its practitioners must maintain an independence from those they cover.
It must serve as an independent monitor of power.
It must provide a forum for public criticism and compromise.
It must strive to make the significant interesting and relevant.
It must keep the news comprehensible and proportional.
Its practitioners must be allowed to exercise their personal conscience.
Citizens, too, have rights and responsibilities when it comes to the news.

one with online access can do just that — generate and publicly disseminate information.

But the Internet does much more than that. The user can include direct links to other information on the Internet, while search engine software automatically scans all web contributions, mapping them against all other similar information and linkages throughout the Web. This is key to how a well-formed search query can lead an Internet user to an otherwise obscure website with just the click of a mouse. The total cycle can be described as a continual process of generate-post-and-automatically-interconnect. This hyper-interconnectivity of the Web makes for its power as an information resource.

The World Bank[14] estimates 2 billion Internet users worldwide at the conclusion of 2010, with Internet saturation in North America and the European Union reaching 70-80% range. This alone represents over 600 million Internet users. (A further analysis of the distribution and demographics of Internet users is beyond the scope of this paper. Still, this information is of interest as it relates to potential target audiences for bio-enterprise-generated online content.)

This base Internet capability, when matched with the size of the Internet community, would be motivation enough; however, newer web tools have emerged which interconnect the Internet users and give them greater roles in creating richer and unprecedented content.

The terms "social media" and "Web 2.0" (pronounced "web-two-point-oh") came into popular use almost simultaneously in the 2004 timeframe. A new wave of web tools had become available, and Internet users could easily publish blogs and respond to comments. There were numerous forums in which Internet users could interact with each other. Early web conferencing services enabling collaboration were only hampered by low bandwidth. The ability to build databases wherein any Internet user could write over another's contribution, improving or correcting it — the so-called "wiki" — enabled the launch of Wikipedia. Massively multiplayer online games were drawing thousands of players simultaneously and online game credits and virtual weapons were being sold on eBay[15], while the online sale of a stolen virtual weapon led to the homicide of a young man in Shanghai.[16] At the same time, the first wave of one-person-to-many-people social networking sites had been launched or at least were in their first embodiments. That would include Friendster (2002), LinkedIn (2003), MySpace (2003), and Facebook (2004).

The concept and actual term "Web 2.0" arose out of a brainstorming session jointly held by O'Reilly Media and MediaLive International leading to the first Web 2.0 conference in October, 2004 in San Francisco.[17] Alternative attributions with respect to the origins of the term itself can be understood when you consider common technical parlance: Each significant upgrade of a technology receives a completely new version number. The term "Web 2.0" was meant to suggest that something new and dramatic was going on with the web. That others may have used the term around or even before that same time merely means that they were discussing significant changes underway on the World Wide Web and understood technical protocol. The conference itself and the entry of the definition of the term "Web 2.0" on the Internet in September, 2005[17] codified the term irrefutably.

To that end, Web 2.0 is meant to represent the technology platform from which social media arises.

There is much discussion as to the precise definition of "social media." If one includes the technology platform itself, then Web 2.0 is a part of social media. If one considers the information generated — the databases containing the user-created content — as the social media, then Web 2.0 is a separate, necessary and supporting entity.

Kaplan and Haelain[18] argue that to understand "social media" you need both concepts: "Social Media is a group of Internet-based applications that build on the ideological and technological foundations of Web 2.0, and that allow the creation and exchange of User Generated Content."

The Directorate for Science, Technology and Industry of the Organization for Economic Co-Operation and Development (OECD) studied user-created content and its potential impact in great detail, resulting in the 2007 report, "Participative Web and User-Created Content: Web 2.0, Wikis and Social Networking."[19] In its view, user-created content is "defined as: *i)* content made publicly available over the Internet, *ii)* which reflects a certain amount of creative effort, and *iii)* which is created outside of professional routines and practices."[19] This "separate[s] it from content produced by commercial or quasi-commercial entities for commercial purposes."[19]

While the message for the bio-enterprise — and for any professional media outlet — is that what they produce is *not* user-created content, social media tools can be provided to target audiences, and their participation and the user-created content which results can be used strategically. In 2010, Genomic Health, Inc. launched an integrated social media campaign entitled "Pass It On ... Until Every Woman Knows."[20] It used Twitter, YouTube and Facebook to "prime" an online community to educate women with early stage invasive breast cancer (and those around them) about the value of using Genomic Health's OncoType DX test for their treatment decision. Housed at www.mybreastcancertreatment.org, it encourages visitors to become members, to "like" regular entries on the Facebook page and leave comments, to share their stories, to view and spread the videos on You-

Tube — separate videos in English and in Spanish, and to "follow the campaign on Twitter and re-tweet messages." Genomic Health's data indicates that in the U.S. use of the test results in less chemotherapy use in 30% of patients, while "only half of the patients who are eligible for Oncotype DX get the test." As one can see, this is a powerful and compelling message, with an online campaign which goes far beyond simply publishing data on a corporate website. Note, however, that this is a bio-enterprise-driven social media campaign, for which the bio-enterprise may also attempt to gain exposure via traditional media outlets in the consumer health and wellness space. These traditional media outlets, while not generally in the science-business space, serve as such since they are delivering a consumer-level message for the mature bio-enterprise. At this level, the need for communicating science within the message itself is minimal; thus, commonly-available public relations strategies directed toward consumer health would apply.

Further note, that with new technology being introduced every day, with the ability to incorporate smart phones, iPads and other personal consumer devices, in addition to personal computers, et al, and with new uses for existing technologies constantly being conceived by the users themselves, a precise definition of social media is not possible — it is and will remain an evolving space.

This is an essential concept for the bio-enterprise as it strategizes its on-line content.

Table 4 presents sample social media categories, along with definitions and examples of their presence on the Internet.

Table 4: Select Social Media Examples

Social Media Category	Definition with Examples
Collaborative Content	Joint and simultaneous creation of content by many end-users Examples: Wikipedia, Wiktionary, WikiTravel
Blogs	Date/time-stamped entries in reverse chronological order with opportunity for comments and interactions Examples: BiotechBlog, BIOtechNow, BusinessInsider
Content Communities	Shared Media Content Examples: YouTube, Flickr
Social Networking Sites	Individuals connecting through personal information and one-on-one/group communications Examples: Facebook, LinkedIn, Google+, eBay, Twitter
Virtual Worlds	Users interact using three-dimensional graphics, either replicating real life or entirely fictional; can be social interactions or can be part of a multi-player game environment Examples: Second Life, War of Worldcraft

Science communication and social media

While continual review of all media coverage for accuracy in science relevant to the bio-enterprise is called for, the greatest potential for unedited, unverified science information lies in the online space, especially in those spaces created by social media tools and reflective of multiple, unassociated authors.

A recent multi-industry study of 1,100 companies, which included focused interviews with over 700 executives directing efforts in the social media space, was published in the July 1, 2011 edition of the *Harvard Business Review.*[21] Asking "What's your social media strategy?", it suggests that at least four approaches have emerged, depending upon organization need. In the case of bio-enterprise, a verification of the coverage of its relevant science is included in these needs, as is what the public considers to be valid sources of scientific information to its liking.

One prime online source is Wikipedia, which provides volunteer-created tutorials and explanations in many areas of science. Wikipedia's own entry entitled "Reliability of Wikipedia"[22] has no fewer than 196 non-Wikipedia sources as of this writing. The early rejection of Wikipedia by traditional journalists and editors, as recently as 2008[23], is independent of how the general population may perceive the validity of the science which appears in Wikipedia. This continues to be an interesting consideration, while the position of these same journalists and editors appears to have softened.

Analyses of an eight-year period from The New York Times, The Washington Post, The Wall Street Journal, USA Today and The Christian Science Monitor actually demonstrated an increase in sourcing material from Wikipedia. In another example of interdependence, it appears that traditional media has in fact begun to validate Wikipedia.[24] Whether this will increase the value of Wikipedia as a dependable scientific source, or it will decrease the value of the traditional media space, remains to be seen.

Regardless, all audience-available explanations of science — independent of pedigree — must be sought out, understood and strategized by the bio-enterprise.

Executive-level communication

Strategically, the bio-enterprise seeks the most impactful media coverage possible, and as documented earlier, research points to sources in traditional journalism as having the greatest impact on the total media space, online included. In the biotechnology industry, this points directly to traditional media outlets and those items placed by science-business journalists.

When examined in terms of interdependence, this captures trusted audiences, while driving factual sources through to non-traditional media outlets. To that end, how to engage science-business journalists is an essential question.

An analysis of traditional media coverage reveals that "CEOs were (the) most commonly used sources for business."[1] In fact, CEOs were the lead source above all others for professional business journalists — some 36% of all business sources.

With respect to whole-organization coverage, professional journalists are generally reluctant to interact with other personnel or spokespeople for the organization.[1] There are numerous explanations for this journalistic practice, but the primary explanation is obvious on its face: The legal responsibility for all aspects of the organization rests on the shoulders of the CEO, while the CEO is also responsible to his or her board of directors. The consequence to the CEO who misrepresents his or her bio-enterprise — either intentionally or inadvertently — is significant in the extreme. These external pressures on the CEO provides the highest level of assurance vis-à-vis the validity of the information provided to the professional journalist.[1]

Since the CEO is the preferred source of information for professional journalists, from a strategy standpoint, executive-level management requires substantive media skills, including recognizing the ongoing need to foster these relationships. From a strategic standpoint, this presents real opportunity, as there is much to be strategically communicated about the bio-enterprise.

At every step in the life of a bio-enterprise — from initial funding through continued scientific research, product development and regulatory acceptance, product launch and full operation, merger or sale and more — the need to reach various target audiences evolves. At the outset, this often includes venture capitalists, industry analysts, and institutional investors, before evolving to such targets as healthcare organizations, insurance providers, advocacy movements, individual consumers, and more.

Concerns to be addressed can be long-term or transient. They can cover an innovative business proposition or a newly-perceived risk in an accepted business proposition, the news of promising, unclear or failed scientific results, the creation/dissolution of alliances and partnerships, and the list goes on. In each instance, communication must be strategized by the bio-enterprise to strategically identify which audiences need to be reached, and what messages need to be delivered.

The result can affect reception in the venture capital community, reaction in public markets, questions in the regulatory environment, the perception of consumers, and even the emergence of a wholly-constructed science controversy, for which there is no scientific basis.[25]

Thus, *pro-active* media engagement is as essential as *reactive* media engagement. To that end, a model is presented under which all such engagement with the global science-business media can be strategically addressed throughout the bio-enterprise life cycle, independent of the demands of pro-active vs. reactive media engagement.

BIO-ENTERPRISE MEDIA STRATEGIES

While successful media strategies intend to influence the media, they are also based on trust. Thus, the model must include all human participants, all media outlets, all audiences, and the trust relationships between them. These relationships may be human-to-human or media-to-human, and they can only operate successfully when trust is understood and honored.

Still, to be motivated in this regard, the bio-enterprise must subscribe to the need for media engagement and a media strategy.

The need for media engagement

While the dynamic of the global media landscape has been discussed as it relates to the bio-enterprise, questions may remain as to need: Why can't the bio-enterprise simply reach the target audiences directly? Why does it need the media? Is there a way for the bio-enterprise to simply ignore or sidestep the media?

The first two questions go directly to the nature of trust. Assertions made directly by the bio-enterprise are seen as serving the purposes of the bio-enterprise and have no third-party verification; all audiences look elsewhere for that trust.

On the third point, the position of simply ignoring the media, the answer lies in the dynamics of trustworthy information. Most individuals relevant to the bio-enterprise today search the Internet, and that cycle is well-established. The most trustworthy information present on the Internet traces its roots back to professional science-business journalists and traditional media outlets. Thus, the most opportunistic target for impacting target audiences is working strategically with trustworthy science-business journalists.

In short, the bio-enterprise which avoids developing a strong science-business media strategy misses its opportunity to influence the total media space which directly impacts it.

Human participants in the model

The human participants in the strategic science-business media model are the bio-enterprise, professional science-business journalists, other trusted science-business journalists, and ultimately, target audiences.

Within the bio-enterprise, this starts with the CEO, but also includes media relations professionals — both internal and external — as well as anyone within the bio-enterprise who may be able to provide first-hand, credible information.

The model identifies professional science-business journalists as specific candidates with whom to develop relationships, and it also identifies trusted *online* science-business journalists. This latter is that subset of all online journalists, who in the experience of the bio-enterprise has displayed trustworthy past performance and a recognized following with desired target audiences.

Qualifying online journalists requires consideration, as simply self-identifying as a journalist does not guarantee that such a contributor subscribes to a professional journalistic code of ethics, or that their media outlet does. The presence or lack of a published code of ethics can be an indicator, as would be the existence of a science–qualified editorial staff, but is not a necessary qualifier. Professional reputation can supplant it. One example would be the long-time professional science journalist who is now independent and writing a science blog. Also, while an online media outlet may be new, the reputation of its journalists may be of longer and well-established origin, as is their following. Every contributor and every media outlet stands alone, and these online journalists must be considered by the bio-enterprise for relevance to its strategy.

This all leads to reaching the desired target audiences with the appropriate information, and is the entire point of the strategy. It is situational in nature, as is the selection of candidate journalists to be approached for any particular strategy.

Media outlets and information in the model

The media outlets in the science-business media model encompass both traditional science-business media outlets and other online content generated by online science-business journalists, as well as information created by the bio-enterprise and all other online content considered to be bio-enterprise-relevant. Although all enterprises generate content in physical form, this is not addressed by this model.

A strategic science-business media model

As with any business strategy, the bio-enterprise needs to understand what it can directly control, and what it cannot control. Within the global media landscape, the bio-enterprise can control two primary spheres: (1) the relationships it attempts to make, and (2) the content it generates. The bio-enterprise uses what it can control to potentially influence and persuade.

Working backwards from any information that the bio-enterprise wishes to reach a target audience, this model seeks to identify all human participants and media outlets in between, as well as their precise interrelationship — whether human-to-human or media-to-human in nature.

Figure 1 contains a graphical portrayal of the strategic science-business media model for the bio-enterprise. How the model elements interrelate and what the bio-enterprise can control goes to the heart of the influence the bio-enterprise may strategically exert. To that end, all elements and relationships (or portions of relationships) in the model which the bio-enterprise can control appear in GREEN, while elements and relationships which the bio-enterprise cannot control appear in BLUE.

One can see from the model that via the online space, it can potentially access by all human participants in the model — from audiences to journalists; the greatest impact and attendant reach, however, comes from its relationships with trusted professional science-business journalists, both in traditional media outlets and in the online space.

What is also important is what the bio-enterprise cannot control. For example, the bio-enterprise can seek to have relationships with various targeted journalists, but it cannot control if the journalist is willing to have a relationship in return. Even when contact is successfully established, the journalist is ethically bound to fact-check all information provided by the bio-enterprise, to attempt to verify that the overall picture is understood, and to present other valid insights and positions. Inde-

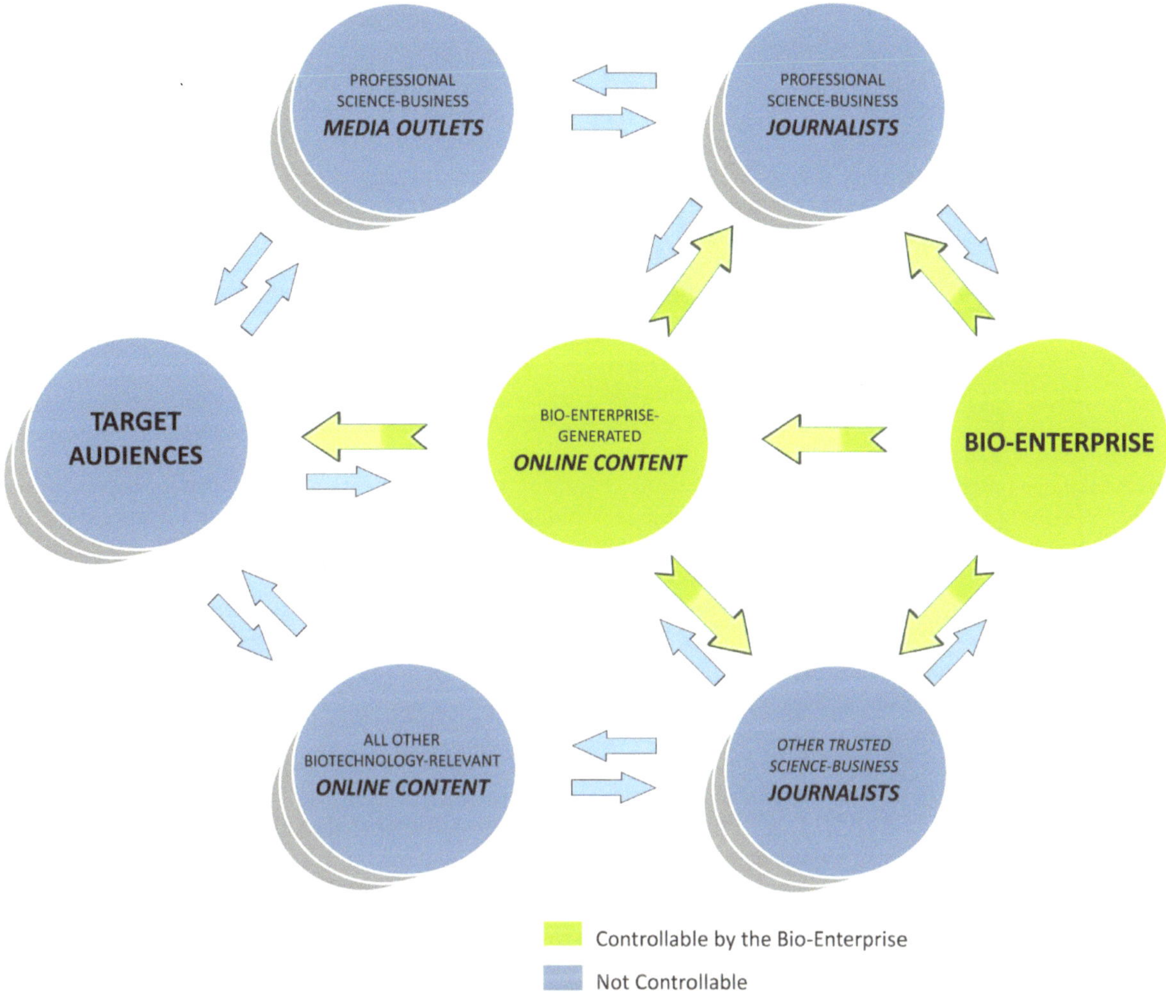

Figure 1: Strategic science-business media model for the bio-enterprise

pendent of whatever information is provided to the journalist — either by formal bio-enterprise press information or via CEO or other human participant — what will actually be published in his or her media outlet is also not controllable, and can often very much less or different from what was provided. Furthermore, the context in which information was presented by the media outlet is uncontrollable. On a final note, the bio-enterprise needs to understand that will be no review prior to publication for professional journalistic media outlets.

Even passively, while the bio-enterprise can control what information appears on its website, it cannot assume that journalists, investors, consumers or any other target audience will read/watch/listen/utilize it. If they do, there is further no guarantee the information will be absorbed, understood and/or subsequently presented to the liking of the bio-enterprise.

The strategic science-business media model simply presents the *potential for influence* on the global media landscape; it does not guarantee it.

Ethical persuasion

In truth, there is every expectation on the part of professional journalists that the bio-enterprise is seeking positive coverage for its organization, often through constructed newsworthy items. In the presence of negative information, the bio-enterprise is expected to attempt to persuade journalists to see the story through its lens.

Still, the more respectful and longstanding the relationship of a bio-enterprise with any journalist and the journalist community, the better potential for successful influence at every juncture. But key to fostering these relationships is understanding and respecting the ethical standards of the journalist, and conducting the bio-enterprise's media relationships in an ethical manner.

Table 5: The TARES Test: Five principles of ethical persuasion[26]

Truthfulness (of the message)
Authenticity (of the persuader)
Respect (for the persuadee)
Equity (of the personal appeal)
Social Responsibility (for the common good)

One approach is for the bio-enterprise to adopt the TARES Test, which was first put forth in 2001 as a guideline for ethical persuasion.[26] Table 5 outlines its five simple principles. Following these or similar principles establishes trust with journalists over time, and enables the bio-enterprise to frame those aspects of the story which might be perceived unfavorably if viewed without the perspective of the bio-enterprise. The deeper the relationship with the journalist, the quicker a response can be put into place.

Should the bio-enterprise choose to adopt what would be considered unethical tactics, such as intentionally creating omissions which the journalist discovers sooner or later, over-reaching, suggesting deliberate misperceptions, etc., there will be consequences. This is a breach of trust.

What is perceived to be unethical persuasion tactics on the part of a bio-enterprise can potentially compromise both the journalist and the media outlet. It can permanently disrupt the relationship between the bio-enterprise and the journalist/media outlet, as well as other journalists and other media outlets. At its worst, unethical practices on the part of the bio-enterprise can become adverse media stories for the bio-enterprise itself — and there are few, if any, professional science-business journalists left with whom to plead its case.

Ensuring Valid Journalist Resources

While journalists are ethically bound to give a total picture of the whole, there are times when that information is simply not available. This is not information specific to the bio-enterprise, but rather the national or global market or backdrop in which it operates.

One example would be the challenge of the journalist attempting to cover global genetically-modified agriculture, which was described in the Columbia Journalism Review.[27] In situations such as these, global data might solely be available from pro-industry organizations, or from advocacy organizations, which have a counterpoint. At times, there may be no science to back up the claims surrounding the available data. This presents an unworkable situation for the professional journalist.

Thus, an innovative media strategy might creatively and transparently supporting independent organizations to establish independent verifiable data, possibly even independent science, which in turn will be rewarded in the long-term maximum success of the bio-enterprise. The key word here is *transparently*, as discovery of the support of the bio-enterprise at a later time will compromise the perceived legitimacy of the information. Strategies which establish consortiums of multiple organizations with counter-balanced agendas would be one vehicle through which acceptable global data can be generated, giving validity to data which positively frames the bio-enterprise and its business proposition.

DISCUSSION

Many elements of the online information space, traditional media, social media, and the interrelation between traditional media and social media are now being studied. Much of the research has focused on topic areas with the greatest volume of information and/or traffic. This includes such subject areas as politics, sports, and business, yet in no case have significant and broad-based media studies been undertaken with respect to the bio-enterprise.

With respect to research which included study of the journalists themselves, two types of professional writers were considered most akin to or inclusive of professional science-business journalists. Distinguished in some studies were popular science writers from general journalists, while in other studies, the business media were considered as a whole.

For the purposes of this article, media phenomena which held true across subject areas were considered relevant to the construction of the model, and as were those media phenomena reflective of general business media.

As mentioned in the general text, much of the science communication work available was concerned with the clear explanation of science to varied audiences. Science communication, strategically focused in service of the bio-enterprise, has not been generally studied, and should not be confused with publications of media advice by media relations professionals. This input is indeed valuable, but should be viewed as advisory by the bio-enterprise, and applied and weighed within the bio-enterprise for its own strategic purposes within the context of the model offered.

Actual future cases of media strategies within the bio-enterprise could prove helpful, although they may not be accessible as the financial considerations, the status of the biotechnology product, the drivers for seeking media attention, et al are often considered confidential, and can be compromising to the current bio-enterprise and its ongoing journalistic relationships.

Certainly, one reliable source which professional science-business journalists repeatedly rely upon is the peer-reviewed science journal. While apparently attractive as it can receive media attention, it was not considered as a strategic media source which can be called for in a timely or controllable manner. When such an article is published, it can be handled by the bio-enterprise under the model as an instance of a newsworthy item.

Finally, the model was specifically developed to withstand the inevitable emergence of new technologies and the continued evolution of media interdependencies.

CONCLUSION

As the bio-enterprise seeks to engender positive reception by various target audiences at every stage in its life cycle, and endeavors to avoid or ameliorate negative coverage or misinformation, the development of an effective media strategy becomes paramount.

With the emergence of social media, the continued expansion of online information, and the need to address science communication with relation to its business proposition, among other biotechnology industry issues, the bio-enterprise can face these challenges in context.

The strategic science-business media model for the bio-enterprise provides a framework against which all media strategies and their respective action plans can be gauged. It shall remain valid in so long as professional science-business journalism continues to inspire trust.

ACKNOWLEDGEMENTS

For their vision and encouragement, thanks are due to Stephen Sammut, Senior Fellow, Wharton Health Care Systems and Entrepreneurial Programs and Venture Partner, Burrill & Company, and Arthur Boni, the John R. Thorne Chair of Entrepreneurship and Director, Donald H. Jones Center for Entrepreneurship, Tepper School of Business, Carnegie Mellon University.

REFERENCES

1. Messner, M. and DiStaso, M.W. (2008) The Source Cycle: How traditional media and weblogs use each other as sources. Journalism Studies 9(3): 447-463.
2. Messner, M. and Garrison, B. (2011) Study shows some blogs affect traditional news media agendas. Newspaper Research Journal 32(3): 112-126.
3. Hong, J. and Sheehy, M. (2010) Growing number of bloggers see their work as journalism. Newspaper Research Journal 31(4): 38-47.
4. American Association for the Advancement of Science (AAAS). (2008) Communicating Science: Tools for Scientists and Engineers. Communicatingscience.aaas.org/Pages/newmain.aspx, accessed 5 November, 2011.
5. Blum, D., Knudson, M., and Henig, R.M. (2005) A Field Guide for Science Writers: The Official Guide of the National Association of Science Writers. Oxford University Press.
6. Condit, C. (2007) How geneticists can help reporters to get their story right. Nature Genetics Review, 8(10): 815-820.
7. Hartz, J. and Chappell, R. (1997) Worlds Apart: How the Distance Between Science and Journalism Threatens America's Future. Nashville, TN, USA: First Amendment Center.
8. Rees, M., et al (2006) Science Communication: Survey of Factors Affecting Science Communication by Scientists and Engineers. London, UK: Royal Society.
9. Peters, H.P., Brossard, D., de Cheveigne, S., Dunwoody, S., Kallfass, M., Miller, S. and Tsuchida, S. (2008) Science Communication: Interactions with the Mass Media. Science, 321(5886): 204-205.
10. Pew Research Center's Project for Excellence in Journalism. (2011) Ethics Codes. www.journalism.org/node/125/print, accessed 5 November, 2011.
11. Kovach, B and Rosenstiel, T. (2007) The Elements of Journalism: What Newspeople Should Know and The Public Should Expect. New York: Three Rivers Press.
12. Committee of Concerned Journalists. (2011) A Statement of Shared Purpose. www.concernedjournalists.org/node/380, accessed 5 November, 2011.
13. Mooney, C. and Kirshenbaum, S. (2009) Unpopular science. The Nation. 17 August: p1.
14. The World Bank. (2011) Data Indicator: Internet users. http://data.worldbank.org/indicator/IT.NET.USER, accessed 23 January 2012.
15. Tyrell, P. (2003) Realities of a virtual economy: INTERNET: Online games are taking on a parallel life of their own - spawning trade, exchange rates and regulatory headaches. Financial Times, 29 December: p8.
16. BBC News (2005) Chinese gamer sentenced to life. 6 August, http://news.bbc.co.uk/go/pr/fr/-/2/hi/technology/4072704.stm, accessed 23 January 2012.
17. O'Reilly, T. (2007) What Is Web 2.0: Design Patterns and Business Models for the Next Generation of Software. Communications & Strategies 65(1) 17-37.
18. Kaplan, A. and Haenlein, M. (2010) Users of the world, unite! The challenges and opportunities of Social Media. Business Horizons. 53(1): 59-68.
19. Organization for Economic Co-Operation and Development (OECD). (2007) Participative Web and User-created Content: Web 2.0, Wikis and Social Networking. ISBN 978-92-64-03746-5.
20. Genomic Health, Inc. (2010) Pass It On ... Until Every Woman Knows. http://www.genomichealth.com/en-us/

Company/Perspectives.aspx?id=7678bf31-6698-4a62-9dfe-33f2f4cdcf28, accessed 23 January, 2012.

21. Wilson, H.J., Guinan, P.J., Parise, S. and Weinberg, B.D. (2011) What's your social media strategy? Harvard Business Review, 89(7/8): 23-25.

22. Wikipedia. (2011) Reliability of Wikipedia. Wikipedia entry, en.wikipedia.org/wiki/Reliability_of_Wikipedia, accessed 5 November, 2011.

23. Shaw, D. (2008) Wikipedia in the Newsroom. American Journalism Review 30(10): 40-45.

24. Messner, M. and South, J. (2011) Legitimizing Wikipedia. Journalism Practice 5(2): 145-160.

25. Ceccarelli, L. (2011) Manufactured Scientific Controversy: Science, Rhetoric and Public Debate. Rhetoric & Public Affairs 14(2): 195–228

26. Baker, S. and Martinson, D.L. (2001) The TARES Test: Five Principles for Ethical Persuasion. Journal of Mass Media Ethics. 16(2&3): 148-175.

27. Gustin, G. (2010) Seeds of Change? Columbia Journalism Review. 48(5): 47-49.

Achieving optimal financial and strategic transaction outcomes for small to mid-sized privately funded start-ups

Received: November 17 2011; revised January 29 2012

Benjamin P Chen
is a scientist/entrepreneur/merchant banker. He is currently the Managing Partner of Ignatius Transaction Partners, LLC - a life sciences merchant bank. Dr. Chen has experience in building talented research teams in immunology, stem cell biology, genomics, gene therapy and molecular diagnostics in both biotechnology start-ups and multinational pharmaceutical companies. He was most recently a Managing Director at Burrill LLC where he assisted a global clientele in completing licensing, partnering and M&A transactions. One of Dr. Chen's passions is to assist life sciences entrepreneurs in developing products that would make a difference for patients in need.

Christa Nicholas
is a Director at Burrill LLC. She has over 20 years experience in marketing, strategic planning and portfolio management in life science companies ranging from start-ups to multinationals. Ms. Nicholas specializes in helping biotech companies optimize the commercial value of their development projects for strategic transactions and/or market entry.

ABSTRACT

Non-dilutive funding and equity capital are two key reasons why life sciences companies pursue strategic partnerships. In fact, alliances are also strong contributors to successful "exits", either through mergers and acquisitions (M&A) or an initial public offering (IPO) associated with market launch. Approximately 40% of partnerships ultimately result in acquisition by the partner. Further, 80% of approved biopharmaceutical products from 2000-2010 had a commercial partner on board[1]. In the current environment, strategic alliances and funding can come from many sources, including the traditional "large pharma" universe — but the question remains: How best for a small management team to gain access to and maximize success with these sources? The focus of this article is to describe how entrepreneurs can leverage external expertise via intermediaries to achieve their near-term and longer-term objectives.

Journal of Commercial Biotechnology (2012) 18, 55–59. doi: 10.5912/jcb.531
Keywords: intermediaries; partnering; financing; emerging markets

INTRODUCTION

Rapid advances in scientific knowledge have created a wealth of innovative biopharmaceutical product opportunities and companies; however raising adequate amounts of Venture Capital (VC) continues to be challenging. Looking forward through the next 3 years, the National Venture Capital Association projects that 40% of VCs will reduce their life sciences investments. Despite solid historical life science portfolio performance (see Figure 1), VCs and their Limited Partners are concerned about recent trends including:

- Increased development times
- Unpredictable FDA approval requirements
- Reimbursement hurdles post-launch
- Limited exit opportunities with an anemic IPO market and acquisitions still relatively infrequent (64 acquisitions of private biopharmaceutical companies took place in 2011)[2]

Correspondence: Benjamin P. Chen, Ignatius Capital Management/Ignatius Transaction Partners, LLC, US. Email: chenbenjamin@yahoo.com

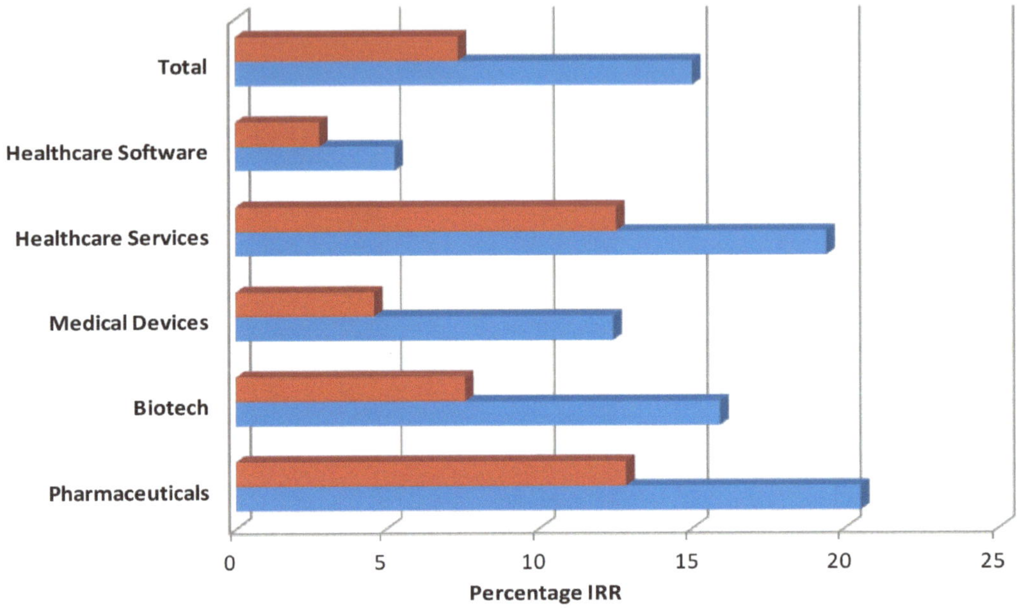

Figure 1: Venture capital returns in life sciences & healthcare, 2000-2010
Source: adapted from NVCA Benchmarking Database

While corporate venture groups are increasingly important in filling the funding gap in early-stage research, it is clear that fewer companies will be able to raise the total amount required to reach an exit from traditional VCs. Even companies backed by strong syndicates are actively seeking alternative capital sources including government, non-profit foundations, and commercial partners in both developed and emerging countries. Each of these funding options has particular issues that need to be considered and addressed.

Grant monies from the NIH, DOD or other US government agencies can be an excellent source of capital — a few companies have bootstrapped themselves all the way through Phase 2b from these sources. However, the amount of government funds may come with strings attached, such as loss of control over IP, choice of lead indication or project timeline. One key strategy is to target ex-US government and sovereign wealth funds, many of which are seeking to establish or grow biotechnology hubs. While these funds may also have limitations by type of technology or the requirement for a local footprint, if interests are aligned, the opportunity can be transformative. China's Sinopharm and Russia's Russnano are just two examples of ex-US sources for funding bioscience technologies.

Traditionally, companies would court investors in the United States and Europe to fund product development to clinical proof-of-concept before exploring opportunities in emerging markets. Today entrepreneurs need to proactively assess the product opportunity globally. Is there significant market potential in a specific geography due to genetic or environmental factors? How do commercial considerations such as pricing policy and distribution affect this potential? Could a partner and/or investors from the local market increase the probability of success or offer a faster route to launch? As shown in Table 1, biopharmaceutical growth opportunities in many emerging markets are in the double digits, while the traditional triad of North America, Western Europe and Japan is growing at less than 5% overall.

The regulatory and reimbursement environments outside the United States may provide an earlier route to market, but both must be considered in making timeline assessments. For example, a highly constrained pricing policy, or one that is reference-based, would be less attractive initially. A dialog with local experts and partners is essential to mapping the most attractive launch/partnering strategy globally.

Non-profit groups such as the Muscular Dystrophy Association, the Cystic Fibrosis Foundation and the Leukemia & Lymphoma Society also support significant earlier-stage R&D. Both inside and outside the US, foundations and angel investors focused on particular diseases offer funding for appropriately aligned projects. In emerging markets, angel investors are more common than foundations. In recent years, angels have become much more thorough in their diligence, deploying expert teams for review and valuation. Leveraging advisors can

Table 1: Pharmaceutical outlook 2011-2015

	Market size (US dollars - billions)	Compound annual growth rate (CAGR)
Global	1,065 – 1,095	3 – 6%
North America	345 – 375	0 – 3%
Western Europe	170 – 200	0 – 3%
Central & Eastern Europe	72 – 82	6 – 9%
Middle East & Africa	35 – 45	7 – 10%
Latin America	72 – 82	11 – 14%
Asia Pacific	195 –225	13 – 16%
Japan	110 – 140	2 – 5%

Source: Burrill & Company; adapted from IMS Health 2011

help gain access to angel capital, based on a match between their investment objectives and a specific product opportunity.

Despite the growing appeal of alternatives, strategic alliances with global pharmaceutical and biotechnology companies still dominate the non-dilutive funding plans of most entrepreneurs. Virtually all clinical-stage companies have assets available for partnering or purchase, however the reality is that only about 5% of these actually result in a significant deal — and the number is continuing to shrink.[3] Why? Firstly, the buyer universe is consolidating (see Figure 2). Every large scale pharma-pharma or pharma-biotech transaction both reduces the number of targets, and alters the business development structure and priorities of the surviving company. Secondly, restructuring continues even in companies without M&A activity, increasing the complexity of decision-making[4]. What then can companies do to leverage themselves into the "5%"?

Proof-of-concept data from a well-constructed Phase IIb study that supports a compelling target product profile is the core value proposition for potential licensees. Equally importantly, the Phase II data must be supplemented with an agreed development plan and regulatory path through pivotal trials to approval, to enable the target opportunity to be assessed in the context of the costs required to reach the market. Some products can be partnered pre-proof-of-concept depending on the nature of the trial, endpoints and target population, and novel platforms often garner early alliances.

Entrepreneurs may have a different view of what constitutes proof-of-concept (PoC) than potential partners. For example, completing a Phase II study showing a statistical difference vs. placebo in the primary endpoint is often described as achieving PoC by the originator. For the partner however, there may be many additional requirements. Regarding trial design and analysis, these could include: 1) primary endpoint selection that is the same as the approvable endpoint in Phase III, 2) use of the final commercial product configuration, 3) at least two dose levels of active drug tested, 4) a comparator arm that represents standard-of-care, 5) statistical power to detect clinically meaningful benefit in the most conservative analyses (worst case imputation), 6) improvement in secondary endpoints, including validated patient reported outcomes, to enhance competitive differentiation, 7) strict limits on acceptable adverse event rates, particularly for chronic therapies.

Beyond the human trial results, partners often expect ancillary studies to be completed, such as long-term safety in two species and a basic drug interaction study, as well as validated manufacturing processes and analytical testing, preferably at a site that has a track record of passing a pre-approval inspection. Of course, a program need not meet every "requirement" to be partnerable. An advisor can help assess the extent to which the data package supports risk mitigation and the resulting valuation implications for the program. In some cases additional work might be recommended, if the time and cost are justified by attaining a meaningfully increased value inflection. Alternatively, an advisor may recommend an option agreement or alternative structure to bridge a temporal valuation gap between the parties.

The quality of the science, IP and development process are key to partnering success — but increasingly, market issues are deemed equally important by many potential licensees. Large partners apply both technical and commercial "probability of success" factors to their Net Present Value (NPV) models, and these factors can often be definitive in determining the attractiveness of an opportunity. Entrepreneurs need to consider how the science translates into addressing an unmet medical need (preferably targeted via a companion diagnostic), in the context of evolving standards of care,

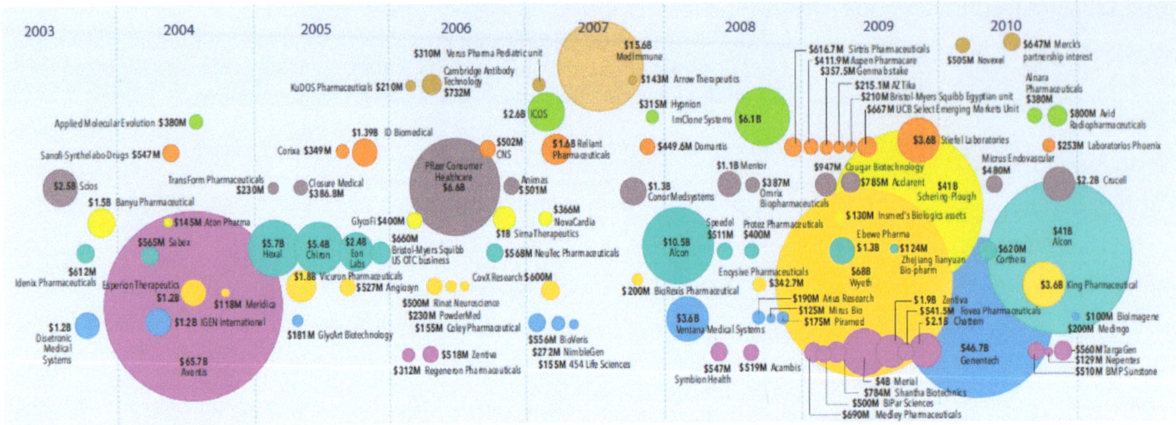

Figure 2: Consolidation of pharmaceutical industry (major acquisitions 1989-2010).
Source: Burrill & Company

competing therapies in development and even medical device and surgical alternatives to drug therapy. Finally, the value proposition of the product should be validated through primary market research with payers, practicing physicians (not just key opinion leaders) and in some cases patients. Data from a well-conducted study adds substantiation to the assumptions that are presented in the originator's demand forecast. Many entrepreneurs assume this work is not needed if it will be duplicated by a partner. However experience has proven that well-designed and executed research provides an important benchmark that accelerates the process of attaining an agreed range for target revenue and profit potential. The same data can be leveraged for other types of fundraising.

From the perspective of the entrepreneur evaluating potential partners, it is helpful to consider the key factors influencing whether the program is ultimately successfully commercialized, as the deal with the highest upfront may not offer the greatest value. External advisors can help assess:

- Is the product/technology strategically central to the partner?
- Does the partner have sufficient domain expertise and marketing prowess in the key markets?
- Is there financial commitment to full development? Would an internal competing program confound decision making?
- How good is the partner's regulatory approval record in the target markets?
- Have the partner's recent alliances been successful? Is there a cultural fit?

Once there is interest from multiple potential partners, the process of translating confidential diligence into term sheets and final agreement begins. This is a long route, averaging 12 months[5], and replete with challenges. Term sheets are generally non-binding and include both the proposed business structure and financial terms. Many entrepreneurs focus on the preliminary upfront payment and milestones, although these may shift in confirmatory diligence. The outline of the transaction structure, in contrast, should remain generally consistent and if properly negotiated, sets the stage for a successful close. It is best to be clear upfront on whether an M&A or partnering deal is the ultimate objective (and which makes the most sense from the buyer/licensee perspective). If partnering is the preferred approach, a further decision point relates to the data package — is there a near term value-creating event that favors an opt-in? After licensing negotiations begin, there are numerous areas that can make or break an alliance:

- Defining the optimal development plan and who is responsible for which aspects
- If planned budgets are exceeded in a co-development structure, what happens?
- What constitutes "commercially reasonable efforts" and how these can be enforced?
- What is the best way to structure a co-promotion arrangement when conventional "detailing" is becoming less relevant to commercial success?
- In the event of a change of control of either party, what happens?
- What triggers termination and what rights survive it?

During negotiations, it is crucial to maintain activity with more than one party for as long as possible, both to increase the probability of bringing the deal to a close and to improve financial leverage. Experienced external advisors can be very useful in providing resources for parallel negotiations and running a synchronized process, supplementing internal business development capabilities.

Given today's highly dynamic transaction environment, it is not surprising that many company executives and boards are availing themselves of external support. This can take various forms, including pre-transaction consulting to optimize the product profile, development plan or regulatory strategy, "introduction" services to link company management with senior contacts at the right targets in the right geographies, and full-service advisors who will lead and manage all aspects of the transaction process. These services are offered by individual consultants, strategic advisory boutiques, and investment banks — the latter offering the ability to include debt or equity financing in conjunction with a strategic transaction. The choice of intermediaries should be made based on project status, transaction objectives, company infrastructure and organizational "fit". Service providers need to work closely with internal staff based on an engagement structure that clearly aligns incentives.

Serial biotechnology entrepreneurs understand that the path to success is full of obstacles, challenges and detours. Ultimately, those entrepreneurs who get rewarded are those who recognize early on the importance of 1) early positioning of the company in a global context, 2) focusing on the advancement of the lead product opportunity and 3) bringing the right external resources, advisors in at the right time.

REFERENCES

1. Deloitte Recap 2011
2. Biocentury January 2012
3. Campbell Alliance Biotech Investors presentation January 2012; Windhover 2012
4. Biotech 2011: The Burrill & Company 25th annual report on the life sciences industry
5. Deloitte Recap 2011

Partnering with the NIH: Now part of the "Value Proposition" for start-ups

Received: October 31 2011; revised January 16 2012

Steven M. Ferguson
is Deputy Director, Licensing & Entrepreneurship, Office of Technology Transfer, National Institutes of Health, Department of Health and Human Services

ABSTRACT

Abstract With its "value proposition" statement a start-up company needs to convince potential investors or pharma partners how it will add more value or solve a problem better than others. High value, low cost assets such as those from the NIH ranging from technology to funding to assistance provide such biomedical firms an excellent jump-start in reaching their goals.

Journal of Commercial Biotechnology (2012) 18, 60–67. doi: 10.5912/jcb.510

INTRODUCTION

For many years the United States has led the world in government funding of non-military research and development (R&D), notably support for basic and clinical research that directly relates to health and human development. A longtime focal point for such federal investments in biomedical research has been the National Institutes of Health (NIH) along with other government laboratories and university-based research programs. Base funding provided by the NIH alone reached $31.2 billion (excluding economic stimulus funds) in fiscal year 2011; approximately 10% of this funding was spent on internal NIH R&D projects (intramural research) carried out by the approximately 6,000 scientists employed by the NIH. The balance was distributed in the form of grants, contracts and fellowships for the research endeavors of 325,000 non-government scientists (extramural research) at 3,000 colleges, universities and research organizations throughout the world.[1] Each year this biomedical research leads to a large variety of novel basic and clinical research discoveries — all of which generally require commercial partners in order to develop them into products for consumer, scientist, physician or patient use. Thus federal laboratories and universities need and actively seek corporate partners or licensees to commercialize their federally-funded research into products in order to help fulfill their fundamental missions in public health.

1 See NIH Overview at http://www.nih.gov/about/

Correspondence: Steven M. Ferguson, Office of Technology Transfer, National Institutes of Health, Department of Health and Human Services, US. E-mail: sf8h@nih.gov

OPPORTUNITIES FOR BIOENTREPRENEURS AND START-UPS AT NIH

With well-established mergers and acquisitions across the entire industry, large consolidated pharmaceutical firms such as Pfizer, Novartis or GSK typically now look for later stage, more mature technologies for in-licensing and further development — not the typical pre-clinical invention arising from traditional research programs at the NIH or at universities. This provides a significantly greater opportunity for entrepreneurs and new companies to step in and fill this gap in the product development by taking on these early technologies from research institutions and bringing them to a stage that is acceptable for acquisition, later-stage clinical trials and marketing by large biotech or pharma companies. The reality now is that commercial partners, especially small, innovative ones, are essential to the role of federally-funded research institutions in delivering novel healthcare products to the market. From new or invigorated activities in technical assistance to express technology licensing agreements, to non-dilutive grant funding, there is an attractive array of available options available from NIH that can be utilized to launch or grow start-up companies. Several of these options will be examined in more detail.

IN-LICENSING OF TECHNOLOGY FROM NIH

As is the case with universities, the NIH cannot commercialize its discoveries even with its considerable size and resources — it relies instead upon partners. Commercializing technologies such as vaccines or drugs and then marketing them successfully in a world-wide market is not the responsibility or mission of research institutions or government agency. Companies with access to the needed expertise and money required are needed to undertake continued development of these inventions from NIH or other research institutions into final products. Typically, a royalty-bearing exclusive license agreement with the right to sublicense is given to a company from NIH (if NIH–owned) or the university (if university-owned) to use patents, materials, or other assets to bring a therapeutic or vaccine product concept to market. Exclusivity is almost always the norm for FDA-regulated products due to the risk involved in time, money and regulatory pathway to companies and their investors. Financial terms of the license agreement are negotiable but due reflect the nascent, high risk nature of the discovery. Because the technologies coming from NIH or NIH-funded research are most typically pre-clinical inventions most licensees are early stage companies or start-ups rather than larger firms who typically want only more proven ideas for new products. In addition to the license agreement there will also often be research collaborations between the licensee and the NIH or university to assist with additional work needed on the product technology. When the licensee is able to sufficiently "de-risked" the technology through its various efforts, these companies then sublicense, partner or get acquired by larger biotech or pharmaceutical firms for the final, most expensive stages of development with the large company expected to be sell the product once it reaches the market.

Since the 1980s federally-funded health research institutions such as the NIH have developed an active but increasingly strategic focus on improving public health through technology transfer activities. As such they are particularly interested in working with start-ups and other early stage companies in the health care area that are looking to develop and deliver innovative products. Rather than just seeking a financial return through revenue generation these institutions are looking to utilize licensing of nascent inventions as a way to increase new company formation, supporting faculty recruitment and retention, enhancing research funding, creating in general a more entrepreneurial culture within the organization, attracting venture investment and development to their specific region (universities) or to the health sector in general (NIH).

ECONOMIC DEVELOPMENT ASPECTS OF LICENSING & TECHNOLOGY TRANSFER

The economic development potential of biomedical research is being recognized as a fourth mission for such institutions -- going along with education, research and public or community service. Thus it is in this "fourth mission" that bioentrepreneurs can play a key role by establishing companies driven by innovative research discoveries.

The economic importance of licensing and technology transfer has become better recognized by research institutions, including the NIH, during the recent recessionary period. For example, the overall product sales of all types by licensees of NIH intramural research is now reported by the NIH Office of Technology Transfer as approximately $6 billion annually, the equivalent of mid-tier Fortune 500 company. Economic development also was the focus of the October 28, 2011 U.S. Presidential Memorandum — "Accelerating Technology Transfer and Commercialization of Federal Research in Support of High-Growth Businesses"[2] . This directive from the White House recognized the economic aspects of innovation and technology transfer for federal research in the way it fuels economic growth as well as creating of new industries, companies, jobs, products and services, and improving the global competitiveness of U.S. industries. The directive requires federal laboratories such as the NIH to support high growth entrepreneurship by increasing the rate of technology transfer and the economic and societal impact from federal R&D investments over a 5-year period. During this period federal laboratories such as the NIH will be (a) establishing goals and measuring progress towards commercialization; (b) streamlining the technology transfer and commercialization processes, especially for licensing, collaborations and grants to small companies; and (c) facilitating commercialization of new technology and formation of new start-up firms through local and regional economic development partnerships.

In addition, many universities and the NIH have set up educational programs that train scientists and engineers to have a greater appreciation as to the importance of commercialization. These include entrepreneurship centers and small business assistance programs at many universities[3], and such things as the "Certificate in Technology Transfer" program given at the Foundation for

2 See http://www.whitehouse.gov/the-press-office/2011/10/28/presidential-memorandum-accelerating-technology-transfer-and-commerciali

3 One such program, for example, is Innovate (http://carey.jhu.edu/our_programs/Innovate/)

Advanced Education in the Sciences (FAES) Graduate School at NIH.[4]

NEW LOW COST START-UP LICENSE AGREEMENTS AT NIH

To better facilitate this "fourth mission" of economic development, the NIH has developed a new short-term Start-Up Exclusive Evaluation License Agreement (Start-up EELA) and a Start-up Exclusive Commercial License Agreement (Start-up ECLA) to facilitate licensing of intramural NIH and Food and Drug Administration (FDA) inventions to early stage companies. These new NIH Start-up Licenses are provided to assist companies that are less than 5 years old, have less than $5M in capital raised, and have fewer than 50 employees obtain an exclusive license from the NIH for a biomedical invention of interest arising from the NIH or FDA. NIH Start-Up Licenses are offered to companies developing drugs, vaccines or therapeutics from NIH or FDA patented or patent pending technologies. The new company must license at least one NIH or FDA-owned U.S. patent and commit to developing a product or service for the U.S. market. The licensee may also obtain in the license related NIH or FDA-owned patents filed in other countries if the company agrees to commercialize products in those countries as well.

Financial terms for the Start-up Licenses are designed with the fiscal realities of small firms in mind and feature either: a one-year exclusive evaluation license with a flat $2,000 execution fee (this license can be later amended to become an exclusive commercialization license) or an immediate exclusive commercialization license. The Start-Up Exclusive Commercial License includes:

- A delayed tiered upfront execution royalty, which would be due to the NIH upon a liquidity event such as an initial public offering (IPO), a merger, a sublicense, an assignment, acquisition by another firm, or a first commercial sale;
- A delayed minimum annual royalty (MAR) or a MAR that is waived if there is a Cooperative Research and Development Agreement with the NIH (or FDA) concerning the development of the licensed technology and providing value comparable to the MAR. Additionally, the MAR will be waived for up to five years during the term of a Small Business Innovation Research (SBIR) or Small Business Technology Transfer (STTR) grant for the development of the licensed technology;
- An initial lower reimbursement rate of patent expenses which increases over time to full reimbursement of expenses tied to the earliest of: a liquidity event, an initial public offering, the grant of a sublicense, a first commercial sale, or upon the third anniversary of the effective date of the agreement;
- Consideration by NIH of all requests from a start-up company to file new or continuing patent applications as long as the company is actively and timely reimbursing patent prosecution expenses;
- A set earned royalty rate of 1.5% on the sale of licensed products;
- A set sublicensing royalty rate of 15% of the other consideration received from the grant of a sublicense;
- Anti-stacking royalty payment license provision can be negotiated by company if it encounters a stacking royalty problem. A stacking royalty problem can occur when a licensee's third party royalty obligations add up to such a high total royalty number such that the project becomes unattractive for investment, sub-licensing or self-development due to low profit margins. Royalty stacking can especially be a problem in the development of biologics due to the breadth of possible third party IP that may be needed compared with traditional small molecule drugs.
- Mutually agreed upon specific benchmarks and performance milestones, which do not require a royalty payment, but rather ensure that the start-up licensee is taking concrete steps toward practical application of the licensed product or process.
- NIH Start-Up Commercial Licenses represent a significant front-end savings in negotiation time and money for new companies since an exclusive license even for an early stage technology might well have expectations prior to negotiations of a immediate execution fee of up to $250,000 or more, a minimum annual royalty due in the first year and beyond of up to $25,000 or more, immediate payment of all past patent expenses and ongoing payments of future patent expenses, benchmark

4 For more details see www.faes.org.

royalties in the range of up to $1,000,000 or more, significant sublicensing consideration and earned royalties in the range up to 5% or more depending on the technology.

Because many, if not most of the technologies developed at the NIH and FDA, are early stage biomedical technologies, the time and development risks to develop a commercial product are high. Depending on the technology and the stage of formation, of the potential licensee company, the company may prefer to enter into the Start-up EELA to evaluate their interest before committing to a longer term Start-up ECLA. Bioentrepreneurs can identify technologies of interest by searching licensing opportunities on the NIH Office of Technology Transfer (OTT) website[5] and by following through with getting in touch with the listed licensing contact. Model template agreements for the Start-Up Licenses and other details on the licensing process can be found on the OTT "Start-up Webpage"[6].

RESEARCH COLLABORATION PROGRAMS AT NIH FOR START-UPS

For some entrepreneurs there is a misperception that NIH scientists (unlike their university counterparts), are not allowed to interact with private sector firms due to the implementation of strict government ethics and conflict of interest rules. While it is true that NIH investigators, in general, cannot engage in outside consulting with biotechnology and pharmaceutical companies in their personal capacity, the fact is that technology transfer-related activities are actually among the "official duties," in which NIH scientists are encouraged to participate. These activities may include the reporting of new inventions from the laboratory and assisting technology transfer staff with patenting, marketing and licensing interactions with companies. NIH scientists can also officially collaborate with industry scientists through the use of various mechanisms including more complex Cooperative Research and Development Agreements (CRADAs) and Clinical Trial Agreements (CTAs) as well as simpler Confidential Disclosure Agreements (CDAs) and Material Transfer Agreements (MTAs).

In a CRADA research project, which could run for several years, NIH and company scientists can engage in mutually beneficial joint research, where each party provides unique resources, skills and funding, and where either partner may not otherwise be able to solely provide all the resources needed for successful completion of the project. In such an arrangement, the details of the research activity to be carried out and the scope of the license options granted to discoveries emanating from the joint research are clearly spelled out in advance. A CTA would typically involve the clinical testing of a private sector company's small molecule compound or biologic drug. The company gains access to the clinical trial infrastructure and clinical expertise available at NIH; however unlike as occurs with a CRADA the company partner does not have any licensing rights to intellectual property that is generated during the clinical research project. NIH usually enters into these agreements only in cases where such trials would be difficult or impossible to run in other places. NIH is particularly interested in clinical trials involving rare or orphan diseases that affect 200,000 or fewer patients per year in the U.S. A Material Transfer Agreement is a popular mechanism for exchanging proprietary research reagents and is used by scientists worldwide. NIH investigators actively use this mechanism to share reagents with scientists in other non-profit organizations. Proprietary and/or unpublished information can be exchanged between NIH researchers and company personnel in advance of making a decision to enter into a CRADA or CTA via the use of a CDA.

Of the collaborative mechanisms described above, a CRADA is perhaps the most comprehensive and far-reaching. Such agreements can provide additional funds for an NIH lab, while providing the collaborating company with preferential access to the NIH scientist's future discoveries and access to scientific and medical expertise during the research or clinical collaboration. A CRADA is not, however, intended to be a means for NIH to provide funding for a new company; in fact, the NIH cannot supply any funding to its CRADA partners. The easiest way for an entrepreneur to access this expertise is to simply approach the agency officially either by contacting a scientist directly or by contacting the institute technology transfer office and/or technology development coordinator[7].

If an early stage company needs access to NIH materials for commercial purposes outside a formal collaboration, this usually would be done utilizing an Internal Commercial Use License Agreement rather than a MTA. These are non-exclusive license agreements to allow a licensee to use (but not sell) technology in its internal programs. Here, materials (either patented or unpatented) are provided, and drug screening uses are permitted. The financial structure of this agreement can be either a single payment, paid-up term license or annual royalty payments, though the second structure is more popular with

5 See http://www.ott.nih.gov/Technologies/AbsSearchBox.aspx

6 See http://www.ott.nih.gov/docs/PHS-Startup-License-Term-Sheet-05172011.docx

7 See http://www.ott.nih.gov/nih_staff/tdc.aspx

start-up companies. Each functions, however, without "reach through" royalty obligations to other products being used or discovered by the licensee. "Reach through" royalty provisions in a license agreement are particularly detrimental to start-up firms as they create downstream royalties or grant-back rights to the licensor on the future sales of downstream products that are discovered or developed through the use of licensed technology, even though the final end product may not contain or otherwise infringe the licensed technology. Popular internal research technologies licensed in this manner include such materials as animal models and receptors.

BASIC & CLINICAL RESEARCH ASSISTANCE

Basic & clinical research assistance from NIH institutes may also be available to companies through specialized services such as drug candidate compound screening and pre-clinical and clinical drug development and testing services, which are offered by several programs. These initiatives are particularly targeted towards developing and enhancing new clinical candidates in the disease or health area of particular focus at various NIH institutes. The largest and perhaps best known programs of these types at NIH are those currently run in the National Cancer Institute (NCI)[8]. The NCI has played an active role in the development of drugs for cancer treatment for over 50 years. This is reflected in the fact that approximately one half of the chemotherapeutic drugs currently used by oncologists for cancer treatment were discovered and/or developed at NCI. The Developmental Therapeutics Program (DTP) promotes all aspects of drug discovery and development before testing in humans (preclinical development), and is a part of the Division of Cancer Treatment and Diagnosis (DCTD). NCI also funds an extensive clinical (human) trials network to ensure that promising agents are tested in humans. NCI's Cancer Therapy Evaluation Program (CTEP), also a part of DCTD, administers clinical drug development. Compounds can enter at any stage of the development process—with either very little or extensive prior testing. Drugs developed through these programs include well-known products such as cisplatin, paclitaxel and fludarabine.

Beginning in 2012 the NIH has been able to established a new center, called the National Center for Advancing Translational Sciences (NCATS), that is designed to assist companies with the many costly, time-consuming bottlenecks exist in translational product development[9]. Working in partnership with both the public and private organizations, NCATS will seek to develop innovative ways to reduce, remove, or bypass such bottlenecks to speed the delivery of new drugs, diagnostics, and medical devices to patients. The Center will not itself be a drug development company, but will focus more on using science to create powerful new tools and technologies that can be adopted widely by translational researchers in all sectors.

NCATS was formed primarily by uniting and re-aligning a variety existing NIH programs that play key roles in translational science. Programs that will be integrated into NCATS include:

- *Bridging Interventional Development Gaps* - which makes available critical resources needed for the development of new therapeutic agents.
- *Clinical and Translational Science Awards* - which fund a national consortium of 60 medical research institutions working together to improve the way clinical and translational research is conducted nationwide. These institutions will serve as a primary test bed for NCATS activities.
- *Cures Acceleration Network* - which enables NCATS to fund research in new and innovative ways.
- *FDA-NIH Regulatory Science* - which is an interagency partnership that aims to accelerate the development and use of better tools, standards and approaches for developing and evaluating diagnostic and therapeutic products.
- *Molecular Libraries* - which is an initiative that provides researchers with access to the large-scale screening capacity necessary to identify compounds that can be used as chemical probes to validate new therapeutic targets.
- *Office of Rare Diseases Research* - which coordinates and supports rare diseases research.
- *Therapeutics for Rare and Neglected Diseases* - which is a program to encourage and speed the development of new drugs for rare and neglected diseases.

There is additional assistance available to firms in other in other disease areas including infectious diseases, drug abuse and many others. A general web portal for listing such public resources has been put together

8 For more information about DTP, see http://dtp.nci.nih.gov/ and for more information about CTEP, see http://ctep.cancer.gov/

9 For the latest developments here, please see http://ncats.nih.gov/

at NIH by the *CTSA (Clinical & Translational Science Awards) Resources for Researchers Webpage*[10]. All in all, such efforts can provide a wide variety of technical assistance (often at little or no cost) for pre-clinical and even clinical development of novel therapies or other biomedical products by start-up firms.

SELLING PRODUCTS TO THE NIH

One of the most commonly overlooked NIH opportunities by biomedical-focused companies is the ability to sell products and services at NIH. Indeed for start-up companies looking to develop new products used in conducting basic or clinical research, the NIH may be their first customer. With an intramural staff of about 18,000 employees, laboratories in several regions of the country (with the Bethesda campus in Maryland home to the majority), and an annual intramural budget of about $3.1 billion, NIH is perhaps the largest individual institutional consumer of bioscience research reagents and instruments in the world. A variety of mechanisms for selling products and services to the NIH are possible, including stocking in government storerooms. Selling to NIH can be a seen as a daunting task for new companies because of the U.S. government's complex acquisition process. However, there are a few simple steps that companies can take, such as establishing a Blanket Purchase Agreement (BPA) with NIH and getting their goods and services into the NIH stockroom. Once these hurdles are cleared, it is much easier for NIH scientists to buy from such companies, and if the quality of goods and services provided by a particular biotech company is superior, an NIH scientist can justify buying solely from that very source.

Companies that provide products and services to NIH laboratories can not only generate cash flow and revenues to fuel R&D, but also begin to demonstrate their commercial acumen to would-be partners and investors. Being a large research organization, the NIH has numerous R&D contracting opportunities. For further information on such opportunities, visit the NIH Office of Acquisition Management and Policy website[11].

The annual NIH Research Festival is also an excellent starting point for companies hoping to sell products to the NIH[12]. This event is held every fall at the Bethesda, MD campus and every spring on the Frederick, MD campus. Part scientific, part social, part informational and part inspirational, this three-day event draws a variety of small to medium-sized bioscience companies. These events attract almost 6,000 NIH scientists, many of whom come to these gatherings to learn about and potentially purchase the latest research tools and services.

NIH FUNDING OPPORTUNITIES FOR START-UPS — SBIR PROGRAM

In addition to contracting opportunities, the NIH can provide private sector entities with non-dilutive funding through the SBIR and STTR programs[13]. The NIH SBIR program is perhaps the most lucrative and stable funding source for new companies and unlike a small business loan, SBIR grant funds do not need to be repaid.

Other noteworthy advantages of SBIR programs for small companies include: retention by the company of any intellectual property rights from the research funding; receipt of early stage funding that doesn't impact stock or shares in any way (e.g., no dilution of capital); national recognition for the firm; verification and visibility for the underlying technology; and finally, generation of a leveraging tool that can attract other funding from venture capital or angel investors.

The SBIR program itself was established in 1982 by the Small Business Innovation Development Act to increase the participation of small, high technology firms in federal research and development activities. Under this program, departments and agencies with R&D budgets of $100 million or more are required to set aside 2.5% of their R&D budgets to sponsor research at small companies. The STTR program was established by the Small Business Technology Transfer Act of 1992 and requires federal agencies with extramural R&D budgets over $1 billion required to administer STTR programs using an annual set-aside of 0.3%. In FY 2010 NIH's combined SBIR and STTR grants totaled over $690 million.

The STTR and SBIR programs are similar in that both seek to increase small business participation and private-sector commercialization of technology developed through federal research and development. The SBIR Program funds early-stage research and development at small businesses. The unique feature of the STTR Program is the requirement for the small business applicant to formally collaborate with a research institution in Phase I and Phase II.

Thus the SBIR and STTR programs differ in two major ways. First, under SBIR program, the principal investigator must have his/her primary employment with the small business concern at the time of award and for the duration of the project period, however, under the STTR program, primary employment is not stipulated. Second,

10 This can be found at: https://www.ctsacentral.org/content/resources-researchers

11 For specific programs see at http://oamp.od.nih.gov

12 See http://web.ncifcrf.gov/events/springfest/2011/ and http://researchfestival.nih.gov/

13 See http://grants.nih.gov/grants/funding/sbirsttr_programs.htm

the STTR program requires research partners at universities and other non-profit research institutions to have a formal collaborative relationship with the small business concern. At least 40% of the STTR research project is to be conducted by the small business concern and at least 30% of the effort is to be conducted by the single, "partnering" research institution.

As a major mechanism at NIH for achieving goals of enhancing public health through the commercialization of new technology, the SBIR and STTR grants present an excellent funding source for start-up and other small biotechnology companies. The NIH SBIR and STTR Programs themselves are structured in three primary phases.

Phase I: The objective of Phase I is to establish the technical merit and feasibility of the proposed research and development efforts and to determine the quality of performance of the small business prior to providing further federal funding in Phase II. Phase I awards are normally $150,000, provided over a period of six months for SBIR and $100,000 over a period of one year for STTR. However, with proper justification, applicants may propose longer periods of time and greater amounts of funds necessary to establish the technical merit and feasibility of the proposed project.

Phase II: The objective of Phase II is to continue the research and development efforts initiated in Phase I. Only Phase I awardees are eligible for a Phase II award. Phase II awards are normally $1 million over two years for SBIR and $750,000 over two years for STTR. However, with proper justification, applicants may propose longer periods of time and greater amounts of funds necessary for completion of the project.

SBIR-TT Phase I & Phase II: Under this new program (SBIR-Technology Transfer or SBIR-TT) undertaken at the National Cancer Institute (NCI) at NIH and in the process of being expanded to other NIH institutes, SBIR Phase I and Phase II awards are given in conjunction with exclusive licenses to underlying background discoveries made by an intramural research laboratory at the institute.

SBIR Phase II Bridge: The NCI SBIR Program has created the Phase II Bridge Award for previously funded NCI SBIR Phase II awardees to continue the next stage of research and development for projects in the areas of cancer therapeutics, imaging technologies, interventional devices, diagnostics and prognostics. The objective of the NCI Phase II Bridge Award is to help address the funding gap that a company may encounter between the end of the Phase II award and the commercialization stage. Budgets up to $1 million in total costs per year and project periods up to three years (a total of $3 million over three years) may be requested from the NCI. To incentivize partnerships between awardees and third-party investors and/or strategic partners, competitive preference and funding priority will be given to applicants that demonstrate the ability to secure substantial independent third-party investor funds (i.e., third-party funds that equal or exceed the requested NCI funds). This funding opportunity is open to current and recently expired NCI SBIR Phase II projects.

Phase III: The objective of Phase III, where appropriate, is for the small business concern to pursue with non-SBIR/STTR funds the commercialization objectives resulting from the Phase I/II research and development activities.

Those who hope to receive an SBIR or STTR grant from the NIH must convince the NIH institute that the proposed research is unique, creates value for the general public at large through advancements in knowledge and treatment of disease and is relevant to the overall goals of NIH. It is important to contact the program officials ahead of time within the particular component of NIH from where funding is sought in order to determine whether the proposed research plan fits these criteria. For start-ups, generally SBIR applications are most successful when they include: an entrepreneur-founder with experience in the field; a highly innovative technical solution to significant clinical need; an end product with significant commercial potential; a technology in need of more feasibility data that the proposed research project would generate; and finally a project that, if successful, would have reduced risk and become more attractive for downstream investment. At NIH, applications are reviewed three times a year. Companies should also be aware that changes for these programs at NIH will be in the works as a result of the recent re-authorization of the programs by Congress.[14]

CONCLUSION — NIH NOW PART OF THE "VALUE PROPOSITION" FOR START-UPS

With its leading edge research and funding programs and focus on the healthcare market, the NIH has a strong record in providing opportunities for private sector entrepreneurs to create both high growth companies and develop profitable medical products. Indeed, a study published in the *New England Journal of Medicine*[15] in 2011 showed the intramural research laboratories at the NIH as by far the largest single non-profit source of new drugs and vaccines approved by the FDA. Clearly this cannot be done without productive partnerships with private industry — past, present and (of course) future. Savvy bio-entrepreneurs and start-up firms can now come to NIH

14 *Ibid.*

15 *N Engl J Med. 2011 Feb 10;364:535-541.*

not only for funding in the form of SBIR grants, but also for product development leads through various licensing and partnership mechanisms. In addition, the intramural NIH laboratories can be seen as an early adopter customer that embraces new biomedical research products as well as a source of expertise, resources and assistance that may not be available elsewhere. Thus entrepreneurs and start-up firms need to fully comprehend, appreciate and utilize the full value that NIH brings to their own work, product development and, of course, to public health.

Licensing, partnering, strategic alliances and university relationships

Received: November 18 2011

Wesley D. Blakeslee

is Executive Director of Johns Hopkins Technology Transfer. He holds an engineering degree (with Honors) from Penn State University, and a law degree (Honors – Order of the Coif), from the University of Maryland School of Law and is a Certified Licensing Professional. Wes is a former engineer and systems analyst with NASA. He has been a Director of a national bank, and served as Director of Computer Development at the University Of Maryland School Of Law where he also taught Computer Law. Wes joined Johns Hopkins in 1999 as Associate General Counsel.

ABSTRACT

The biopharmaceutical industry has been undergoing change for a number of years and that change is accelerating. Larger pharmaceutical companies are acquiring smaller ones, companies are merging, laboratories are being closed, and the number of scientists performing research in the pharmaceutical industry is declining. Overall, commercial industry, including the biotechnology industry, is becoming more interested in the benefits of collaboration with research institutions.

Universities are also changing their view of relationships with industry. Shrinking federal budgets are causing universities to look at other sources of revenue, including collaborations with industry. Federal and state governments are also looking closely at the benefits of sponsoring university research, and in particular are seeking to accelerate commercialization of university discoveries not only to obtain the benefit of invested research dollars, but also for economic development and job growth. Universities, and in particular university technology transfer offices, must understand these changes and adapt to them.

This paper discusses the university/industry relationships, and the particular issues important to universities which shape that interface.

Journal of Commercial Biotechnology (2012) 18, 68–71. doi: 10.5912/jcb.500
Keywords: university; technology transfer; industry; commercialization; collaboration

INTRODUCTION

There has been a significant change occurring in the biopharmaceutical industry during the past decade which has been recently accelerating and will continue into the future.[1] Biotech companies are merging. Larger companies are acquiring smaller ones. Existing companies are closing laboratories, reducing the number of internal researchers doing basic research.[2] As a result, university technology transfer has also changed. Fewer researchers in the biotechnology industry means fewer scientists reading scientific journals and attending conferences. That results in fewer opportunities for university technology to be "discovered" by pharma through the interaction of science and scientists. Technology transfer has always been a contact sport, as Jane Muir, University of Florida, has often stated. It is certainly now becoming much more so.

Pharmaceutical companies also are looking for later stage, "de-risked" technologies, much farther down the development pipeline than the typical university invention. The era when university inventions were licensed easily to large established companies has come to a close. Universities are much more engaged in dealing with small companies and startups than in years past.

Although large companies are acquiring later stage technologies for development, targeted development of

1 Littman, Bruce H. and Marincola, Francesco M., (2011) Editorial: Create A Translational Medicine Knowledge Repository. Journal of Translational Medicine 9:56

2 The New Economy (2010) Big Pharma's Stalled R & D Machine, 16 June, http://www.theneweconomy.com/science-and-technology/pharmaceuticals/542-big-pharmas-stalled-rd-machine/, accessed October 31, 2011.

Correspondence: Wesley Daniel Blakeslee, Johns Hopkins University, US. E-mail: starman@jhu.edu

early stage technologies remains necessary and desirable. With fewer researchers engaged internally in basic science research, biotech and pharmaceutical companies are increasingly seeking to partner with universities to advance research in areas of interest.[3] We believe that in the future the most valuable university technologies and deals will come from collaborations between industry and academia.

At Johns Hopkins Technology Transfer we have been remaking the way we do business in order to respond to these changes. We are now the business concierge for Johns Hopkins. We are not just about tech transfer, but about building the other relationships that we must have with industry. We exhibit and present at conferences such as Bio, and a number of others, because there is a fundamental change in the way we must do business with industry. So far we have been rewarded for that strategy. We have had five record breaking years in a row, even though these last three years have been difficult economically.

The next hurdle is how to change the whole University. Universities, especially those such as Johns Hopkins that have for years been exquisite bases of independent science, must understand better how to work with industry. In the past, many research universities, including Hopkins, have eschewed industry relationships. That is changing in our university and many others. Some have been quicker to do so. Those whose research is focused primarily on the physical sciences have tended to be more directly involved with industry focused research than the schools engaged in basic biomedical science. However, we are learning how to do collaborative agreements with industry and form such industry relationships. In doing so, we must navigate adverse pressure from within and outside of the university.

Some recent regulations on conflicts of interest, for example, have caused a number of universities to adopt internal policies that prohibit their clinical researchers from even having a cup of coffee with a representative from the pharmaceutical industry. Whether the people who wrote those regulations intended to have that kind of direct consequence is not clear. But we all need a better understanding that industry is not an evil term and that it is not comprised of evil-doers who want to corrupt our research. Rather, commercial enterprises are very important partners in what we all do, especially in trying to move products to the marketplace, which is of course the focus of a technology transfer office.

We must continue to respond to industry changes and how that changes our relationship with industry. I think we are doing a decent job university-wide, as are our sister universities around the country, as we learn better how to engage with our industry partners in order to move inventions through, get them out and get them commercialized. It does no good to invent the cure for cancer if you leave it on the lab bench. Without an industry partner to take that invention out, develop it, and finally get it into a patient, we have not cured anyone. We understand that and we know we must do a better job of finding those industry partners.

To do a better job we need a better understanding of industry and, equally, we need to improve industry's understanding of what we do and who we are. Universities have some special issues that we must deal with and if industry understands those issues, then industry can better understand some of the legal provisions we must have in an agreement as well as some others that we cannot grant even though they seem so reasonable to industry.

The first thing that is different about universities is that in the university space, the money received for an invention flows to a number of partners. Table 1 shows the Hopkins breakdown; most universities are very similar.

Out of gross revenues, we first distribute to other institutions any share they may be due. Out-of-pocket expenses, primarily patent and legal costs, are paid next and then we retain 15% toward the expenses of the technology transfer office. The remainder is distributed as shown in the above chart. The inventors receive 35%, paid directly to them. They also receive, and many of them prize this more than their personal share, 15% for their research budget which helps to fund their research projects. The inventor's department receives 15%, 5% goes to the university central administration, and 30% goes to the inventor's school.

Thus, while we own the intellectual property, the income stream is shared and so there are many partners in the income who have a say in what we do and have an interest in how we do it.

That prevents us from doing a number of things. We cannot, for example, include future inventions in a license. Our inventors typically continue to work on and advance the intellectual property. A common request from industry is to roll new technology into the initial license. However, we cannot agree to that. One of the reasons we cannot is the inventor's share. An inventor of new technology is entitled to share the revenues. If we receive $100,000 as an up-front license fee and share that with the inventors on the original license, we cannot roll in the new invention with new inventors who do not then get a share in that original license fee. In addition, an annual payment to the original inventors would thereafter be reduced by adding in the new technology. Thus we must do a new license. Industry often does not understand that. In industry the invention typically belongs to the institution. There is usually no income sharing with the inventors. Industry can bundle together multiple

3 Kling, Jim (2011) Biotechs Follow Big Pharma Lead Back Into Academia. Nature Biotechnology 29: Pages: 555–556

Table 1: Distribution of licensing revenues

Inventor's Personal Share	Inventor's Research Account	Inventor's Department	Inventor's School	University Administration
35%	15%	15%	30%	5%

technologies and do whatever is needed to get the best value for the institution, including bundling in some inventions at no additional cost to the licensee. Universities must think about the best value for the individual inventors because the inventors get a direct share.

University technology transfer is a balancing process between the goal of faculty service, the desire to advance as many technologies as possible, and the practical consideration of financial responsibilities and funding of the technology transfer effort. Most inventors are not focused on cost, either for the office, filing of patents or other costs incurred, because they are focused on the mission, and expect the university to have a good technology transfer office that gives them very good service. On the other hand, the entities that pay the bills, the individual schools at JHU, want us to be good stewards of the money they give us and they want to see a good return on their investment. We must balance among the missions of advancing technology, making our faculty happy and doing what the schools are willing to fund.

Most university technology transfer offices are cost centers. Many expend more on costs than gross revenues received. Some, like Johns Hopkins Technology Transfer, make an "operating profit", such that gross income exceeds total costs. Still, they are a cost center because of income distribution. The total cost of Johns Hopkins Technology Transfer which is paid by the schools, exceeds the 30% of revenue received by the schools.

While we would like to earn a significant profit for the university, our primary mission is to bring the benefits of discovery to the world. Because of this mission, we always look at who is going to be the best partner. It is not always the biggest upfront dollar figure. The Bayh-Dole Act requires us to ensure that federally funded inventions are developed. We cannot license technology to be put on the shelf by a licensee, where development of the invention could cannibalize the market share of an existing product. We must put the invention with someone who is going to develop it into a product. In so doing, we also give consideration to assuring that essential medicines be widely and readily available.

We also look for a licensee who will have a good relationship with our faculty member. Is there sponsored research that will come in? Does the faculty member have a relationship with the scientist at the company that will be developing this product? Our faculty members have influence in where their inventions go and that is very important to them.

Sponsored research is also a very important component. With the change in structure of the biopharmaceutical industry, there are fewer internal industry researchers, and so there is a growing interest to have the scientist who invented the product continue the research and move it downstream. Most of the things that we license are at an extremely early stage and so need continuing basic research. Universities are the best place to conduct such early stage research and so we look for sponsored research in any deal we do.

We also look at industry-faculty relationships. Many faculty members like to consult and faculty consulting is a way that industry can take advantage of the expertise at universities. Joint development deals are another means of sharing expertise between universities and industry. We are relatively new in that area compared to some other schools but we have a number of those in the works right now and several have been signed recently. We believe there are more such agreements to come as pharmaceuticals look toward universities to assist product development.

We have always done collaborative research, where members of our faculty are working with scientists at other universities or in private industry. No money changes hands, but people work on things together. Intellectual property issues complicate such arrangements. However, contrary to the opinion of industry held by some academics, we have not found the companies difficult to deal with. Also contrary to the opinions of many in industry, universities understand industry needs and are also not difficult to deal with. We are both continuing to learn how to streamline the contracting process. These new relationships need care and attention, but they are very doable.

Conflict of interest and conflict of commitment are important issues to universities. We must be sure that any science we do and report is independent and accurate. We must be sure that if the researcher publishes a paper on the efficacy of a product, the independence of that research is not questioned because the faculty member has a relationship with the company or the university will stand to benefit from the sale of that product.

When industry pays the university to do research, we get reimbursed only for our actual cost. That is how we charge. Universities charge the direct cost of the people doing the work plus a small amount of overhead to cover the actual indirect costs of supporting the researchers. Therefore, we must retain title to any developed intellectual property. We are more than happy to license developed intellectual property to the research sponsor on terms to be negotiated when the intellectual property is disclosed.

The law impacts the flexibility of private non-profit universities to deal with intellectual property. If we were to license in advance any intellectual property that we develop, the sponsored research would be deemed commercial work. The tax and other issues that arise make that very difficult for universities.[4]

Publication is another potential stumbling block in university/industry relationships. Universities must be able to publish the results of their work. We cannot agree to do research in secret. However, we have significant experience in dealing with this issue, and a workable solution can always be found. We can and do often agree to delay publication if needed to protect intellectual property or permit preparation of a patent filing. We do not desire nor is it necessary to publish company confidential information. Often the things important to industry are not the central issue in the science, and so a compromise can be reached, permitting the university researcher to publish the important science, while protecting material of importance to industry.

One thing that is very important to universities is risk aversion. The university does not produce the product. The licensee develops, tests, markets and sells the product. We have no control of that process and so the university cannot be responsible for product injuries to third parties. Thus, the university requires indemnification against product liability in any license agreement. That is an area that gets much attention during license negotiation. We want protection from claims even if there is an allegation that we were negligent because the small amount we receive from a license does not cover the cost of litigation, let alone the risk of damages.

In conclusion, universities are easy to deal with, understand the needs of industry, and are seeking to work with industry. Universities have certain issues and requirements. Once industry understands and appreciates the ways that universities differ from industry, reaching an agreement is usually straightforward and not that difficult. University/Industry collaborations are not only mutually beneficial, but of benefit to society as well. We have different but complementary strengths and expertise, which we must be able and willing to combine if we are to maximize the benefits of university research and truly, as is the mission of Johns Hopkins, bring the benefits of discovery to the world.

---

4 *See, e.g.*, Rev. Proc. 2007 – 47.

What every biotechnology entrepreneur needs to know about VC due diligence

Received: November 30 2011; revised December 5 2011

Stephen M. Sammut
is Senior Fellow and Lecturer, Health Care Management, Wharton School and Venture Partner, Burrill & Company, a venture capital fund focused on the life sciences. During his 20 years on the Wharton faculty, he has created nine courses that he has taught to over 6000 students. During his private sector career, Steve has been involved in the creation or funding of nearly 40 biotechnology and information technology companies globally. He is on numerous Boards including HealthRight International, and the Center for Medicine in the Public Interest. He is a member of the editorial board of the *Journal of Commercial Biotechnology.*

ABSTRACT

Due diligence, as it applies to venture capital, is actually imprecise. Origins of the term are based in banking case law. Due diligence to the attorney is more of a precise concept. A better term is "homework." Better indeed, because the burden of this homework weighs far more heavily on the entrepreneur than on the venture capitalist. The odds of getting funded by a venture capital firm are somewhere between 50: and 100: 1. In most instances, the funding goes to companies that already have some connection into the community of venture capital funds. Does this mean that all others need not apply? That is hardly the case. Good venture capitalists know a good opportunity when they see it, but sometimes it is not always obvious. Either the business plan is flawed in strategy, format or content, or the due diligence process reveals a team totally unprepared to fulfill their own vision. Elsewhere in this special edition of JCB is an article on the business plan and the pitch book. This article teaches, in a manner of speaking, how business plans get read and pitches gets heard.

Journal of Commercial Biotechnology (2012) 18, 72–77. doi: 10.5912/jcb.533
Keywords: due diligence, venture capital; business plan

INTRODUCTION

The entrepreneur starts the due diligence process before the business plan is written. In fact, it should start with the very conception of the business. The entrepreneur's concept for the business must immediately fit into the context of what is already underway scientifically, clinically and commercially, as well as who the players are at each of these levels, including the sources of financing, be they governmental, angel or venture capital. If this sounds like laying the groundwork for competitive analysis, it is. Even before the globalization of biotechnology with some 10,000 companies worldwide today in various stages of development, there were thousands in the United States alone, each building on a platform of intellectual property directed through a development program to meet an unmet clinical need.

Must the VC be told the context of the business? Given that specialty VC funds see over 100 plans per month and do a good job of cataloging these and keeping abreast of developments across several fields, they are concerned that the entrepreneur knows the context of their own business. And on that basis, why that particular VC was targeted to receive the plan. Is the business in a field where the VC has demonstrated some interest? Are there complementary businesses in the VCs portfolio? Are there competitive businesses in the VC's portfolio? On the basis of background and board memberships was a particular person in the firm targeted to receive the plan?

Does this sound like the business plan has to be written for a particular audience? It should because that is the case. That does not suggest that there is a separate business plan for each venture capitalist, but it does mean that the entrepreneurs must learn the VC mindset as it relates to the particular field that the company

Correspondence: Stephen M. Sammut, Wharton School, University of Pennsylvania, US. E-mail: smsammut@wharton.upenn.edu pursues. If a company is directing RNAi technology to an orphan disease, the entrepreneur should search the websites of active VCs to determine who has invested in similar categories, determine when and try to learn why (or why not).

There are a number of other axioms that entrepreneurs have to bear in mind:

- Diligence is an art and not a science; more judgment is applied than method.
- VCs expect that while a business plan must be confident, it is also tentative
- Time frame of diligence has been compressed. This is not a reflection of competition for deals, but a desire to make decisions and move on to the next opportunity.
- A few VCs use a "checklist," but most VC's rely on building an experiential inventory over time. One way of looking at diligence is that it is a recapitulation of everything that has gone wrong in prior investments.
- Beyond the first cut, diligence is less a "yes" or "no" process, and more a means of learning what will have to be managed.
- Opportunities are often turned down for reasons that have little or nothing to do with the merits of the venture itself. Think of these as any biologist would, as exogenous reasons. These include the stage of development of the company relative to where a fund happens to be in its own life cycle. For example, by the fourth or fifth year of a fund's life, an early stage deal may simply not fit a portfolio.
- The reasons for denial of a deal that truly matter are the endogenous ones; some of these will be the focus of the remainder of this paper.

From here, this paper explores the function of a business plan by asking provocative questions about business concepts, establishment of objectives, self-assessment of the technology, the market and financial needs. In addition, this paper also emphasizes the management composition, milestones that management must meet, the opportunity from a prospective corporate partner's point of view, and finally the critical dimension: capitalization needs and strategy. Just like anything under the sun, a business plan has both function and structure, and it is function that that drives the structure.

The function of a business plan is to build a consensus, first among the management team, and then between the team and its customers and investors.

The best plans establish specific criteria, and then do a self-check to make sure that the concept and the opportunity truly line-up. The combination of criteria with analysis forms a self or "auto" due diligence. The exercise ultimately lays open the opportunity, its components and its human, technological and financial requirements. This all adds up to a fancy way of saying: "Oh, this is what we'll need to get going." This seems like a simple statement for the most complex part of capitalizing a new business. It is critical because professional investors must be confident that management understands the nature of the challenge. Moreover, investors want to know how management expects to sequence and breakdown the tasks at hand, and what the associated coats will be. The launch strategy proceeds from the above.

As stated, the primary function is to build consensus. This obvious concept is too often overlooked. Unless management has thought through—as a team—how the opportunity will be attacked, time and money will be lost—and competitors may get to the finish line first.

All parties present must be involved. The leader of the team must drive each of the factors identified here:

- Cohesiveness—unified action (not necessarily full agreement)
- Flexibility—adaptability to circumstances and ideas
- Competence—control of the facts, the team and the situation
- Relevance—that their background, style and personality fit the situation
- Reasonableness—related to flexibility, but with the added willingness to explore all reasonable ideas
- Maturity—adaptability to change and the legitimate needs of all stakeholders, and
- Drive—nothing ever goes as planned. Entrepreneurs need high energy levels and the capacity to try, try, try again.

And all of the above results in a plan organized to fulfill these functions.

As a conceptual guide, if the message to be conveyed takes into account the due diligence process that the investors will follow, the plan becomes persuasive. This process of self-examination should be the basis for each facet: the business concept; the proposed execution; the in-person pitch and presentation; and, the path towards a deal.

THE PROCESS OF "AUTO-DILIGENCE"

Auto-diligence, once again, is thinking about the business from an investor's point of view. We will review the particular inventory of issues, but we are not talking about one set of issues for the entrepreneur and another for the investors. For the most part, each issue is the same—perhaps with two sides. Experience counts for a lot in this regard. Few investors and fewer entrepreneurs work from a checklist. With experience, the issues all merge together, and the process becomes intuitive. For newcomers to the process, however, it is a good idea to take an iterative approach. Some of the issues will be silent, others will be the basis for profound discussion.

This approach sets up a construct to pick apart the due diligence process. This is how the construct works. For each of the major areas: the business concept, plan objectives, on through the list to capitalization planning, we will look at **the ideas** behind them, what is involved in **characterizing the idea**, what the **generic issues** might be, and how to get to the hard-core **specific issues**. This thought framework is applied to each of the following:

- Business Concept
- Capitalization Planning
- Technology/Intellectual Property Assessment
- Market Assessment
- Financial Assessment
- Management Team
- Milestones and Capital Needs
- Corporate partner assessment

By way of example, a few of the above will be examined in the framework of: the underlying **idea** of the category, the **characterization** of the category, **generic issues**, then **specific issues**.

THE BUSINESS CONCEPT

What is the **idea** behind the business concept? Business concepts typically start with the identification of a need. Alternatively, the basis for a business may be a new technology, or better stated, a new technological solution to a known problem. Most businesses fall into one of the above two categories. From time to time, we see businesses that change the world. Use existing companies in the same or similar industry as a frame of reference: the technology base, operating strategy, products services, and so forth. How will these influence the thinking about the ideas behind the concept?

How is the business concept **characterized**? The statement counts. A business concept makes no sense unless it is described within the context of an existing market, base of or way of doing business. A common error is that business concepts are often stated in a way that makes no sense to the person reading or listening. A translation to the tangible is critical.

The next part of the construct is to consider the **generic issues** surrounding the business concept. Either the investor or management can consider: assisting the company in better understanding and articulating their business concept; in characterizing the concept according to market needs that could be filled; technological matters; and, the nature of any breakthroughs. What role can different parties play, e.g., investors, corporate partners, etc. in assisting customers or clients in defining a business concept?

In particular, these sources can describe or validate: market needs filled; technological problems addressed; the nature of the breakthrough; in placing the concept within the framework of regional infrastructure and trends; and, in placing the concept in the context of the industry as a whole

Finally, let's look at the **specific issues** behind the business concept. Precision becomes important at this stage. Ultimately, a professional investor's interest in supporting a customer or client is to accelerate the relationship. Context, as it relates to an investment team's own abilities, reach and technologies is the focus. Among the specific matters to be explored when assessing the business concept are: existing technology base; current operating strategy; current product or service portfolio; possible/probable future strategic direction; competitive environment; and, current and future customer base. The emphasis is the basis of the value-added participation of the investment team.

CAPITALIZATION PLANNING

Our framework now moves into the dimension of capitalization planning. We will look here at **the ideas** behind the capitalization planning, the **characterization**, **generic** and **specific** issues. Capital made available by investors will depend on the nature and stage of business, the scale and scope of the business, the pace at which it will develop, its size, needs (and timing of those needs), the use of capital, the sources, and the instruments that will be used in receiving the capital. The sources and amounts of capital depend on: nature and stage of business; scale of business; pace at which technology develops and products are adopted; relative size of business; what the business needs, how much and when; what it is needed for; debt, equity, government grants or a mix; and the type of securities through which capital flows.

How is capital planning **characterized**? Investors have to know how they can recover their position before they get into a deal. As a consequence, characterization of capital planning should go as far as describing certain event, such as an initial public offering, or sale of the company to a third party. This exit planning, in the early stages of a company's life, will also drive strategy. Bear in mind some principles when characterizing capitalization: sources of capital are based on nature and stage of business, and the pace and magnitude of capital requirements; major expenditures, such as licenses, can be leveraged; needs and returns are characterized based on available exits for investors; and, exit mode often drives strategic planning

There are **generic** implications of capitalization strategy: For what does management build? What are the operating implications? What are the alliance implications? Previously, in this issue of JCB, we reviewed the impact of capital staging on ownership and dilution. Management is always cognizant of this dynamic because it determines their ultimate wealth. Management has to trade off cheap money that adds no additional value, with receiving money from an investor that can make a difference in the outcome of a company. So then, in capitalization planning what are the generic issues: build for IPO? Build to be acquired? Build to make acquisitions? Use milestones and capital forecast to structure stages? Raise capital from the optimal, most value-added source at each stage? There must also be consideration of future capitalization events, and associated requirements of sources of that capital.

Finally, there are critical, **specific** issues for consideration with respect to capitalization planning. Different investors, based on the size of their own fund, their investment expectations, and time frame will accept or reject a business opportunity. The art is to fit a company's profile with a given set of investor's criteria. Understanding what a company will have to surrender for each increment of capital is a part of planning. It means that companies will have to carefully identify and describe the risk factors and how to manage them.

In the article in this issue on legal issues in capitalization, there is a description of preferred stock. Typically, the fine points of these instruments will not affect the company. However, professional investors will have expectations about how the financing of a company was structured in earlier rounds. Forecasting needs emphasizes what works best at each stage, but it is based on the anticipation of future capital events.

Here is a litany of **specific** capitalization planning issues: Do the company's capital requirements fit the criteria of available sources? What will the company have to surrender in the way of ownership given the risk profile? What securities instruments and related terms will be offered? What are the implications of these terms? What alternatives does the company have? How can these be managed? Are they compatible?

TECHNOLOGY AND INTELLECTUAL PROPERTY ASSESSMENT

The framework applies handily to Technology and Intellectual Property Assessment. What is the **idea** of technology? The function of a technology is to make life easier, facilitate tasks, expand productivity and refine efficiency. The benefits of a technology must surpass its costs of adaptation, implementation and use. Technologies, therefore, must either fit naturally into existing systems, or create unprecedented possibilities that redefine means of meeting goals and building value

Characterization of technologies is done by comparisons and analysis. Look for fit within existing systems, the value add to the system, and so forth down the list. Analyzing the technology in these terms provides most of the information and insight needed to characterize a technology, its role in the business, its value to the customers and so forth. Technologies are measured in terms of: their fit with existing systems; value to the existing system; probability of development; risk they represent; their contribution to sustainable competitive advantage; proprietary dimensions; and, their role in the value chain.

The generic issues regarding technology assessment start to be fun. Thinking through how the technology is embodied, how it will be deployed, its functionality and so forth are engineering issues and business issues. They speak not only to production, but to adoptability, and ultimately value. The big picture, of course, requires description and measurement of the developmental risks and constraints, the total costs of implementation and deployment, and duplicability. Here is a list of **generic** issues:

- How is the technology embodied?
- How is the technology deployed?
- What is the functionality: Within a system? As a product?
- What are the development risks and constraints?
- What are the total costs of deployment?
- Can it be duplicated?

The **specific** issues are challenging but enjoyable. Determination of fit into an industry's technology base and capability is an obvious primary issue. Related to this issue is a consideration of the impact of the technology on the customer. An examination of these issues addresses costs and the nature of the risks associated with the technology. The acid test is an assessment of where the customer derives value, and how the company can capture its fair share of that value.

- How will technology fit into the industry?
- How will it affect the way customer operates?
- How will it affect the way the industry operates?
- What are specific costs? Over what time period?
- What are the potential hidden or indirect costs?
- What is really being placed at risk?
- Where will the customer derive value? How will the company and its partners capture it?

MARKET ASSESSMENT

Perhaps the most complex and perhaps elusive application of our construct is to the assessment of the market. What are the fundamental **ideas** that will need to be tested and later characterized? Recall that markets are defined by segment but within each segment are defined: product/service line width; features, functionality; service, availability, image and reputation; selling and relationships; and, price.

The market is specifically **characterized** on the basis of: its **P**osition relative to competition; **P**erception relative to competition; **P**romotion relative to competition; **P**ricing relative to competition. The emphasis on competition is unrelenting and absolute.

What is the overall business impact? Risk? There tends not to be much difference between the **generic** issues and **specific** issues in market assessment. The big picture focuses on two points: performance results and marketplace attributes.

As the above questions are answered by an entrepreneur, direction for the business becomes obvious. If the answers do not come easily, or if they are vague, the business is either non-viable, or management needs a new point of view. Sometimes, markets simply are not ready. Some generic issues, therefore, are:

- What are the key marketplace metrics?
- How do these metrics compare with the competitor's prior performance and stated goals?
- How do these metrics compare to those of other rivals or the local firm?
- Marketplace strategy/attributes
- Strategic coherence
- Balance of potential and gain
- If "second in," what differentiates?
- Reasonableness of assumptions

The specific issues again stress competitive issues. While generic issues address viability and strategy, the specific issues most likely will suggest tactics. As you consider each of the above questions in connection with a proposed business concept, the persuasive points will emerge for the written business plan. Obviously, these specific issues are situationally related. Here are specific issues that an entrepreneur must consider:

- How do the competitor's scope and posture compare with the focal opportunity?
- What customer based advantage does the competitor possess compared with the focal opportunity and vice versa?
- What are the commonalties and differences in goal structure?
- What issues will confront the competitor?
- How will it respond?
- What is the impact on our strategy? What do we need to know?

An example presented at the 2011 Boot Camp by Amita Shukla of New Enterprise Associates puts a finer point on specific issues. When describing the market for, say, a new therapeutic, an entrepreneur might make this representation:

> *There are one million patients worldwide. $5,000 is the cost of drug/patient/year.*

At 80% peak penetration, t**here is a $4 billion market potential.** Well, that sounds all well and good, but the venture capitalist is looking for different filters. Her reasoning might go something like this:

> *There are 1 million patients worldwide but only 400,000 patients are in the US. Only half of these are at the relevant disease stage and of this only half have access. At $5000 per patient per year in the US (and by the way $2500 in the rest of the world) with 50% peak penetration the market in the US is $250 million. Such an assessment presents a more realistic and analytical view of the market potential for the product(s) and their competition* **and *demonstrates an understanding of the markets and competitive dynamics.***

MILESTONES AND CAPITAL

The last example of the use of the due diligence framework will be applied to Milestones and Capital Planning. Strategy for raising capital depends on how much capital is needed, and at what intervals. Capital needs are driven by carefully composed, tested and plotted milestones. Capital should be raised in stages as defined and driven by critical milestones. Linking milestones and capital needs is a fundamental approach to managing risk and accelerating value. Milestones and capital needs cannot be de-coupled, and the way to characterize the two are together. The advice given here stresses several basic concepts.

When characterizing milestones and capital needs, establish the assumptions and critical accomplishments required for progress. Develop a logical progression of critical success factors, demonstrate their linkage. Create a project management chart linking activities, resources and required cash. Develop a capitalization schedule and related budget and plot capital requirements realistically

The generic issues echo the previous examples. It may seem merely cosmetic, but it is important to develop a sense of appropriate level of detail for each of the milestones, their components and accompanying capital requirements. It is not unusual to see budgets for office supplies to be broken down to include paper clips, and in the same plan to see a one line entry for "product development." Think about the non-verbal message that that communicates. The critical, most sensitive components require the greater levels of detail; the care with which these are described and the way that the business is staged inspires the most confidence. Here is a checklist:

- Is the detail appropriate to the tasks?
- If a prospective strategic alliance, is the project management approach consistent with the partner's?
- Have defaults been built in?
- Have slack variables, or failures been built in?
- Is the development scheme the product of consensus?
- Who did the linkages of milestones and capital needs?

The specific issues, once again, get down to tactics. They provide management, and therefore prospective investors, with a handle on assessing the likely costs, and the outcome of spending invested funds. These questions, at their heart, extract the true components of how risk is being managed by the management team. Here is what the venture capitalist will ask and, hence, what the entrepreneurs must ask themselves:

- Is the core technology adequately staged?
- Are lead applications appropriately selected?
- How does the development of the lead applications relate to the core technology?
- Are regulatory or licensing hurdles built in?
- To what degree will acceleration or delays in licensing impacts the end result?
- Are activities appropriately coordinated with partner's activity?

For the purposes of building and critiquing a plan, the criteria, questions and answers are presented.

TYING IT ALL TOGETHER...

Think function first. Prioritize the message on the basis of function. Develop the structure of the plan according to function. State the plan, but build in answers to the essential questions, integrate the information. If the team is unproven, build confidence with clarity, certainty and a drive to consensus. Know your audience by knowing what they have done, how they do it, their biases, their goals, and their style. And never stop asking questions about the business.

Valuation methods in early-stage biotechnology enterprises: The "Venture Capital Method" at work

Received: November 30 2011; revised December 5 2011

Stephen M. Sammut
is Senior Fellow and Lecturer, Health Care Management, Wharton School and Venture Partner, Burrill & Company, a venture capital fund focused on the life sciences. During his 20 years on the Wharton faculty, he has created nine courses that he has taught to over 6000 students. During his private sector career, Steve has been involved in the creation or funding of nearly 40 biotechnology and information technology companies globally. He is on numerous Boards including HealthRight International, and the Center for Medicine in the Public Interest. He is a member of the editorial board of the *Journal of Commercial Biotechnology*.

ABSTRACT

Valuation approaches to biotechnology companies by angel investors and venture capitalists often appear to the entrepreneurs to be based on voodoo rather than sound principles of finance. While there may be some truth to that perception, there is actually a very sound, somewhat complex, internal logic to the way *private* biotechnology companies are valued. Note the emphasis on the word "private." This article will focus entirely on the valuation methodology of companies that are not yet listed. Moreover, the article will not consider the valuation approaches used by, say, a pharmaceutical company when it is targeting a biotechnology company for acquisition or strategic partnering. The emphasis of this piece is on the thought process or algorithm in play — the so-called "Venture Capital Method" — when an investor is looking at a biotechnology company in the seed stage or in the first or second round of venture capital funding.

Journal of Commercial Biotechnology (2012) 18, 78–82. doi: 10.5912/jcb.494
Keywords: venture capital; valuation methods; capitalization

INTRODUCTION

Is the valuation approach to early stage biotechnology companies different from the way a company in another technology sector is valued? Yes and no. The elements and machinations are the same, but the consideration of the end point or terminal value differs. What is terminal value? The terminal value of a company is a determination of what a company will be worth as a whole at the time that the investors can sell their shareholdings to a third party, be that the public or an acquiring company at some arbitrary time in the future. The voodoo essentially starts at this endpoint. First of all, in any technology enterprise how can the future be forecast with anything approaching certainty? It can not. In the case of Google, for example, how could its role in the Internet be foreseen, let alone how that role would be valued upon the Initial Public Offering and thereafter? Google, in hindsight, represented a radical approach to search engines upon which a whole series of underlying business and revenue models were built. Likewise, how could the financial and investment destiny of social media companies be forecast when they were still a twinkle in the eyes of the entrepreneurs long before anyone else had thought about the category? Again, they could not. The companies that are revolutionary serve the purpose of providing vision and hope, and the fuel for entrepreneurial efforts and investment risk-taking. They are not valuable as benchmarks.

Terminal values are determined conservatively and across most industries arbitrarily. In biotechnology, since the exits for investors typically occur post-acquisition (and companies are more often acquired after they have been traded publicly), the investors will use the prevailing "mood" and values at acquisition as a basis of terminal value. BUT, that value is discounted back to allow for time – five to seven years – and risk. The general pattern of these acquisitions, especially once discount-

Correspondence: Stephen M. Sammut, Wharton School, University of Pennsylvania, US. E-mail: smsammut@wharton.upenn.edu

ed, is somewhat less than inspiring. The history of IPOs does not help the situation. Companies that are traded oftentimes are listed at market capitalizations well below $200 million, and their aftermarket values often fall below that. IPOs are fund raising events; they are not exits. Hold that thought: Terminal values are estimates based on conservative interpretations of prevailing acquisition activity with discounting for time and risk. For illustrative purposes in this article, we will work with a "universal" terminal value of $250 million after discounting for market forces and risk. There is, however, a further discounting, as we will see, when allowing for dilutive events.

Once we have a terminal value, there are other critical considerations. The first is the analysis of capital needs. There is an aggregate capital need for any company from the time it gets started to the time, once again, that the investors can exit. Capital needs will vary widely for a therapeutics company (and then by therapeutic indication), a diagnostics company, a prosthetics company, a device company, etc. Suffice it to say, the aggregate number is important, but the staged use of capital as driven by scientific, clinical development and commercial milestones is the critical item to forecast. Experienced venture capitalists generally know what these costs should be for any given set of milestones for any given category of company. The challenge for the entrepreneur is to get these milestones and their cost estimates as close to the norm as possible. For illustrative purposes, let's imagine a specialty therapeutics company that will have an aggregate need of $100 million in capital in order to get up to and through Phase 3A of clinical trials. We will also assume that upon completion of Phase 3A it gets acquired at our previously posited terminal value of $250 million, a figure that has already been discounted.

What happens now? Let's take a look at milestones:

- Financing Round "A" is preclinical, intellectual property, core management team build out: $10 million to be spent over a one year period.
- Financing round "B" is Phase 1: $10 million to be spent over a one-year period.
- Financing Round "C" is Phase 2 A: $10 million to be spent over a one year period
- Financing round "D" is Phase 2B: $ 20 million to be spent over a one year period
- Financing Round "E" is Phase 3A: $50 million to be spent over a one year period.

Note that this is a highly oversimplified and unrealistically symmetrical scenario that is crafted for illustration.

In the aggregate, this company will use $100 million to get to and through Phase 3A. At this point, we have said, it will be acquired for $250 million. More often than not in the real world the company would go public at this point, perhaps to raise sufficient capital for phase 3B, or might enter a strategic alliance. For simplicity, we will ignore these more likely events and go right to an acquisition.

If you have not yet read the article by Ashley Stevens on capitalization tables in this issue of the *Journal of Commercial Biotechnology*, please do so now. From this point forward the article assumes that the reader s familiar with the concepts of:

- Capital structure of the company, i.e., who owns what and through what types of securities instruments
- Dilution, i.e., the impact on ownership percentage of new capital infusions into a company

THE PROBLEM TO BE SOLVED

The situation is that the reader is a prospective investor in the first round and is planning to lead the first round of DNA Therapeutics, Inc. For the sake of simplicity - and this never happens in reality — you are making your calculations based on the common stock equivalents of the securities that you plan to purchase.

Your task as the prospective investor is to answer this question on behalf of yourself and other venture funds that might also participate in this "A" round of financing:

How much ownership in percentage and shares of DNA Therapeutics do we need to own today in order to meet our return expectations at the time of exit?

You are working with the following assumptions:

1. The expected rate of return is 50% per year. We are going to use this expected rate of return as our discount rate. Those readers with a background in finance have probably just let out a loud gasp. In venture capital the unknowns are so great that traditional methods of deriving a discount rate do not function well. The expected rate of return becomes the proxy.
2. We will sell our shares at year five. It would be nice in the real world if we had that kind of

liquidity. Like everything else in this exercise, this is simply an assumption.

3. There will be four additional rounds of financing prior to my exit. Each round of financing will have a dilutive impact of 20% over the previous round. This uniform rate of dilution is irrespective of the amount of money going into the company in those future rounds. We are assuming that the company is on target at the end of each round and has added enough value that the dilutive impact of the new money is a relatively modest 20 %. This does not reflect reality all that well.
4. The current ownership of the company and a stock option pool (the entrepreneurs and future management) is 2 million shares.

APPLYING THE VENTURE CAPITAL METHOD: A SIX STEP APPROACH

STEP 1: Determine the company's terminal value.

In this case we are positing $250 million

STEP 2: Calculate Discounted Terminal Value (DTV)

DNA Therapeutics DTV = Terminal value divided by present value of the future income streams over the holding period. Simply put:

$$DTV = \frac{\$250\ million}{(1 + 50\%)^5} = \frac{\$250\ million}{1.5^5}$$

$$DTV = \frac{\$250\ million}{7.593}$$

DTV = $32, 925,000 – Yes, that works out to a large discount!

STEP 3: Required ownership at exit assuming no additional financing

To calculate the Required Final % ownership (RFP) at the Exit Date:

Investment: $10 million

DTV = $32, 925,000

RFP = Investment/DTV
= 0.304 or 30.4%

Step 4 A: Determine the Number of New Shares

Existing shares: 2,000,000

RFP = 0.304 or 30.4%

New Shares = (2,000,000 / (1-.304) - 2,000,000)

= (2,000,000/.696) - 2,000,000

= 2,873,563 – 2,000,000

= 873,563 new shares

STEP 4B: Allowing for additional rounds

Calculate the retention ratio (the percentage retained after four subsequent rounds each having a dilutive impact, we are assuming, of 20%)

Retention Ratio = (1/ (1.2) / (1.2) / (1.2) / (1.2)
= 48.2 % or 0.482

Calculate the Required Current Percent Ownership on Day 1. This % accounts for how much of the equity is retained, *after* subsequent rounds have been awarded.

The Required Current Percent Ownership

= 0.304 divided by the retention ratio (0.482)

= 0.631 or 63.1%

Step 4C: Shares needed allowing for further dilution

The dilutive impact must be translated into a number of shares:

Original shares = 2,000,000

Revised RFP = 63.1%

New Shares with dilution

= (2,000,000 / (1.00-.631)) - 2,000,000

= 5,420,054 – 2,000,000

= 3,420,054 new shares to be issued

Step 5A: Share price with no future dilution

For purposes of comparison, let's calculate what the share price would be if there were no dilutive impact of future rounds of financing:

Total amount to be invested = $10,000,000

Number of new shares with no dilution = 873,563

$10,000,000 / 873,563 = **$11.45 per new share**

Step 5B: Share price with full dilution

When we allow for the impact of dilution, naturally the share price will fall; in this case, quite dramatically.

Price per share with dilution:

Total amount to be invested = $10,000,000

Number of new shares with dilution = 3,420,054

Price per share = $10,000,000 / 3,420,054
= **$2.95 per new share**

STEP 6A: Pre-money and post-money value of the company allowing for no dilution

The pre-money value is simply the implied value of the company before the new capital is infused. Paradoxically, it is the post-money that is calculated first because, as you will read below, it sets the upward bounds of what might be acceptable in the next round of financing. The Pre-money = Post money – new capital. Obviously, the Pre-Money + new capital = Post money.

Existing number of shares outstanding times the price per new share.

= 2,000,000 shares x $11.45/new share

= 22,900,000 pre-money value, i.e., the value of the company before the new funds are invested.

The Post money = $32,900,000 (the pre-money + the amount invested)

This, again, is the value when we do not take dilution into account.

STEP 6B: Pre-money and post-money value of the company allowing for full dilution over subsequent four rounds

Total number of existing shares outstanding times the price per new share:

Pre-money = 2,000,000 shares x 2.95 (fully diluted)

= $5,900,000

Post-money = $15,900,000

This is a fourfold drop in the pre-money value; dilution counts.

COMMENTS ON SENSITIVITY

The astute reader — even without a background in finance — probably has an intuitive sense of just how sensitive the calculations will be to the key variables.

Terminal value was here set somewhat arbitrarily — but historically supportable — at $250 million. What would have been the pre-money value of the company had it been $500 million. It would have doubled preserving for the founders and entrepreneurs tremendous value.

In like manner, the discount rate applied was 50%. Variations in discount rate have a disproportionate impact, all the more when the time period varies; the impact one way or the other is essentially geometric.

The numbers put to work are all prospective and there tends to be little room for negotiation. The venture capitalist will use a lower discount rate in later rounds of financing or in those cases where significant risk has been reduced. If the market, be it M&A or public offerings is frothy, the venture capitalist may run the numbers with a shorter number of years, but the tendency is to be conservative. Similarly, if IPO values are running higher than the historical average for the sub-sector of the company, the venture fund may be willing to work with a higher terminal value. Here again, investors will tend towards conservative forecasts.

CONCLUSIONS TO BE DRAWN

Magically, the bottom line of this exercise with the simplified, hypothetical numbers actually came out to a fairly conventional valuation, at least in the "A" Round. A pre-money of $5 million to $6 million when the company needs $10 million is not far off the mark. Interestingly, the key number here is the Post-Money value of $15,900,000. Basically, the company will not be on target unless there is confidence at the end of the depletion of the proceeds of the first round has achieved this value and then some. If market conditions are good, a "ramp-up" of value such that another $10 million in the "B" round causes only another 20% dilution would be generally considered a good outcome. The venture fund leading the B round will repeat the exercise that we have done above with new assumptions, as would the leaders of the subsequent rounds.

There are some caveats here; part of the assumption was that all rounds of financing would be computed with common-stock equivalents. In real life, the venture capitalists would use Preferred Stock which has many different rights. Be aware that rights such as "participation and dividends can have profound negative impact on the holdings of the entrepreneurs. For convenience, when we allowed for a 20% dilution we assumed that the impact of these was built into the pricing. When preparing for a negotiation or in assessing the economics of a term sheet, the entrepreneur should factor these rights into the calculations using Ashley Stevens' capitalization tables article as a guide.

Valuation exercises almost always produce a contentious result. The entrepreneur has to bear in mind that the bargain with the venture capitalist is one that buys into a "system" of doing things. The high discount rate of 50% used in the calculations produces a phenomenal return in the face of success, but two-thirds of the investments made are total losses or barely return capital. The take away is that the entrepreneur, for better or for worse, is part of an investment portfolio. The only way to win the game is to stay true to the milestones and projected capital requirements. In such manner, the portion of ownership retained by the entrepreneur will have real value at the end of the game.

The Art of the Cap Table

Received: November 30 2011; revised March 1 2012

Ashley John Stevens
is Lecturer, Strategy and Innovation Department, Boston University School of Management.

ABSTRACT

This article provides an overview of the impact of raising capital on the equity ownership structure of a biotechnology company. The equity ownership structure as captured in a table of capitalization ("Cap Table") determines how the fruits of success will be divided between founders, management and investors at an exit event such as an acquisition or initial public offering. The evolution of the Cap Table is captured and described through multiple financing events and scenarios and illustrates how value is allocated to the various parties involved in the transactions as the company grows and develops.

Journal of Commercial Biotechnology (2012) 18, 83–97. doi: 10.5912/jcb.522
Keywords: venture capital; finance; venture capital; angel investors, IPO; capitalization table; equity; common stock; preferred stock

INTRODUCTION

Founders stock is the first stock issued by a new company to those who found it. It is called common stock since the stock has no special rights or preferences — all shares are treated equally. When a company is first incorporated, it has no assets, has a great deal of technical, team and market risk, and hence has very little value. Therefore, the founders' stock is generally sold at its par value (a nominal value printed on the share certificates) of 1¢ or even 0.1¢ per share. The founders of the company will buy the stock from the company in the percentages they've agreed that each would own of the company. The parties, based on their past and expected future contributions to the company, negotiate these percentages. Despite the low price, if, say, 10 million shares were issued to the founders at 0.1¢, the proceeds to the company would be $10,000. This initial capital should be enough to pay the initial legal fees to incorporate the company, set up employment agreements with the founders, etc.

The founders are free to agree on any distribution of ownership they wish. An approach that will maximize teamwork and camaraderie will be to have equal shares, but there may be significant differences in contribution (e.g., bringing IP to the company, providing initial operating funds, etc.), experience, employment circumstances, duration of their planned employment with the company, and so forth that might dictate a different arrangement.

Correspondence: Ashley John Stevens, Boston University School of Management, US. E-mail: astevens@bu.edu

Many companies are incorporated in Delaware, even if their operations are initially going to be in one of the other 50 states, because of the favorable body of corporate law in Delaware. Lawyers and venture capitalists like to deal with a good understanding of how agreements are enforced in a court of law in the event of a problem. Therefore, venture capitalists will normally insist that companies they are going to invest in be incorporated in Delaware, so it is not a bad plan to incorporate there initially.

One of the quirks of Delaware law is that a company's state taxes depend in part on the number of shares the company has issued and outstanding. In order to minimize the tax bite in the early days of a company, entrepreneurs frequently issue a relatively small number of shares upon founding, and then split or reapportion them when it is time to bring in capital financing.

All employees who receive stock in a company, but particularly the founders because of the large amount of stock they receive, should be required to "earn in" their stock by maintaining their employment with the company for a defined period. This is referred to as vesting. Four years is a typical vesting period for founder/employee stock, with perhaps 5 or 10% vesting immediately, and the remainder over time accordingly to an agreed upon monthly or quarterly schedule. That said, to maximize the tax treatment of their stock, the founders will normally buy all their stock up front and the company will have the right to buy the stock back at the same price the founders paid, with the number of shares subject to this buy back decreasing over time. This is called an 83(b) election. A founder who is irrevocably assigning intellectual property (IP) to the company may be exempted from part or all of the vesting requirement since the company will now have control of the IP going forward irrespective of the future employment of the founder.

Since many biotechnology companies originate based on university research, we will use as an example a university spin-out company, founded by:

- A professor, who is not planning on leaving the university and joining the company, but who will chair the scientific advisory board and consult for the company for the one day per week that academic employment contracts generally permit;
- Two post-doctoral fellows who worked on the technology in the professor's laboratory, are co-inventors with the professor on the patent applications the university filed on the technology, and who will join the company as chief scientific officer and chief technology officer;
- A CEO, who has resigned from a position as vice president for business development of a major pharmaceutical company; and
- The University, which, while not actually a founder of the company, has agreed to exclusively license the professor's technology to the company and has agreed to accept founders' stock in lieu of a cash license fee.

The founders agree that the professor will get 20%, the CEO 40%, the postdocs 10% each and the university 20% of the founders' stock. The company is incorporated in Delaware, so the company sells a total of 10,000 shares to the founders, at a par value of $1/share.

The Cap Table of the company at the end of the Founders Round is shown in Table 1.

The professor owns 20% of the company and his stake is valued at the price that he paid for it, $1,000.

Table 1: Cap Table after Founders' Round

Price per share: $1.00

	Shares	Raised	%	Value
Professor	2,000	$2,000	20%	$2,000
Postdoc A	1,000	$1,000	10%	$1,000
Postdoc B	1,000	$1,000	10%	$1,000
University	2,000	$2,000	20%	$2,000
CEO	4,000	$4,000	40%	$4,000
Total	10,000	$10,000	100%	$10,000
Issued and outstanding	10,000			
Fully diluted	10,000			
Raised in this round	$10,000			
Cumulative raised	$10,000			

THE SEED ROUND

Once the company is founded, the management team agrees that they need to perform some proof-of-concept experiments before the company can approach venture capitalists for a major financing. They decide to approach their friends and family for funding, plus the CEO agrees to invest. They decide they need to raise $200,000 to do this work.

To get the price per share in the Seed Round to be around $1/share, they first split the shares 250 for 1, so the professor now has 500,000 shares and everyone else is increased proportionately. The company now has a total of 2.5 million shares issued and outstanding.

The company decides to sell 250,000 shares at $0.80/share, raising $200,000.

The value of the company before the financing (the "pre-money value") was $2 million (2.5 million shares each worth $0.80/share), while the value of the company after the financing (the "post-money value") is $2.2 million (the $2 million pre-money value plus the $200,000 raised).

The Cap Table after the Seed Round is shown in Table 2.

As a result of the transaction, the founders are all diluted by about 10%, so the professor now owns 18.2% of the company and the "seed investors" own 9.1% of the company. However, the value of the professor's stake has gone from $2,000 to $400,000, so he is not complaining. We're still dealing with common stock at this point.

While in this illustration common shares were issued to the company via a "priced round", it is more common that the company wouldn't actually issue shares to the seed investors, but would issue them convertible notes. In a convertible debt financing the money is borrowed with a promise to repay it, or if certain conditions were met, such as the raising of a Series A Round within a specified time period, to issue shares instead of repaying the loan in cash. It is very difficult to establish the value of a company at the seed stage as we have indicated above, and the result may be a contentious negotiation — not a good thing. If the seed round investors are *friends and family*, they may be unable to place a realistic value on the company. Since the company and the investors probably would not have even agreed on the price of the shares, they agree to leave as part of the terms of the convertible note that the conversion price will be decided by the Series A Round investors. To reward these seed investors for the "use of their money" the price of

Table 2: Cap Table after Seed Round

Price per share $0.80

Split: 250 for 1

	Shares	Raised	%	Value
Professor	500,000		18.2%	$400,000
Postdoc A	250,000		9.1%	$200,000
Postdoc B	250,000		9.1%	$200,000
University	500,000		18.2%	$400,000
CEO	1,000,000		36.4%	$800,000
Seed investors	250,000	$200,000	9.1%	$200,000
Total	2,750,000	$200,000	100.0%	$2,200,000
Issued and outstanding	2,750,000			
Fully diluted	2,750,000			
Raised in this round	$200,000			
Cumulative raised	$210,000			
Pre-Money	$2,000,000			
Post-Money	$2,200,000			

the Seed Round shares will is generally be less than the price of the Series A shares — either by specifying that they will be converted at a lower price per share than the Series A shares, as a discount or by issuing the Seed Round investors warrants to purchase additional shares. In this illustration, the Seed Round investors agree to a 20% discount to the Series A, which is a common level of discount, which will give them a 25% profit when the Series A Round is raised.

SERIES A VENTURE FINANCING

A good outcome is that the proof-of-concept experiments funded by the Seed Round investors are successful and the company decides it is now ready raise its Series A financing withfrom a professionally managed venture capital fund. It decides it needs to raise $3 million to develop its initial product. Two venture funds agree to invest $1.5 million each by buying 1.5 million shares at $1.0/share. They are not prepared to buy common stock but insist on buying a new class of shares, participating preferred, or participating convertible preferred shares. These shares are a type of preferred stock that gives the holder the right to receive dividends equal to the normally specified rate that preferred dividends receive as well as an additional dividend based on some predetermined conditions, such as an acquisition or liquidation (this might be a full return of their capital or with some multiple). Furthermore, in the event of a liquidation or acquisition, the participating preferred shareholders can also have the right to receive the price of their shares as indicated, as well as a pro rata share of any remaining proceeds that the common shareholders receive. Basically they get paid back their investment and then also share in the proceeds with the common shareholders. We'll illustrate this point later in the article.

The issuance of preferred shares at $1.0/share increases the fair market value of the common shares. They will be worth less than the value of the preferred, because of the various preferences that the preferred shares enjoy, but the value will be substantially higher than the par value which the founders paid[1]. If the company were to issue common shares to new employees, they would have to pay income tax on the fair market value of those shares. Therefore, the new investors also agree to allow the company to issue 1 million shares of common stock into an option pool that will issue stock options to new employees that will be hired and paid from the Series A financing, to ensure that the new employees have a financial incentive to see the company succeed. New employees are issued options, not shares, because they would have no way of selling any of the shares to raise the money to pay the ordinary income tax they would owe on the share issuance. An option allows them to get all the benefits if the company is successful without any of the risk if the company is unsuccessful and its shares never achieve any value. There is also a vesting schedule for the shares issued under the option pool.

The Cap Table after the Series A financing is shown in Table 3.

The various shareholders' ownership share of the company now depends on whether the shares from the option pool are included in the calculation or not. The VCs own 52.2% of the shares that are issued and outstanding, a majority, though this will go down to 44.4% when all the options are exercised, i.e. on a fully diluted basis. The professor's ownership share of the company has gone down from 18.2% after the Seed Round to 8.7% of the shares that are issued and outstanding and to 7.4% on a fully diluted basis. However, the value of the professor's shares has gone up a further 25%, to $500,000, so again, he is not complaining and his wife is beginning to think it is worth him being gone that much of the time.

1 The fair market value of the common shares is determined by the board of directors of the company. In the early days of the company, the fair market value of the common shares will probably be about 25% of the value of the preferred. As the company develops, the fair market value of the common shares gets closer and closer to that of the preferred, and has to reach 90% of the value of the preferred 30 days prior to the company's initial public offering. The fair market value is used only for purchases and sales of common shares, not for valuing the company.

Table 3: Cap Table after Series A Round

Price per share: $1.00

	Shares			Raised	%		Value
	Common		Series A		I&O	FD	
	Shares	Options					
Professor	500,000				8.7%	7.4%	$500,000
Postdoc A	250,000				4.3%	3.7%	$250,000
Postdoc B	250,000				4.3%	3.7%	$250,000
University	500,000				8.7%	7.4%	$500,000
CEO	1,000,000				17.4%	14.8%	$1,000,000
Seed investors	250,000				4.3%	3.7%	$250,000
Management Pool		1,000,000				14.8%	$1,000,000
VC Fund A			1,500,000	$1,500,000	26.1%	22.2%	$1,500,000
VC Fund B			1,500,000	$1,500,000	26.1%	22.2%	$1,500,000
Total	2,750,000	1,000,000	3,000,000	$3,000,000	100.0%	100.0%	$6,750,000
Issued and outstanding	5,750,000						
Fully diluted	6,750,000						
Raised in this round	$3,000,000						
Cumulative raised	$3,210,000						
Pre-Money	$3,750,000						
Post-Money	$6,750,000						

The pre-money value of the company was $3.75 million, while the post-money value is $6.75 million.

SERIES B FINANCING

With its product successfully developed and tested and its value proposition supported by hard facts, the company is ready to gear up to have its product manufactured, then to introduce and sell the product to customers. Bringing products to market is an expensive activity, and the company decides it needs to raise $10 million, and because of the great data from testing the product, it is able to justify a doubling of the share price, to $2/share. The two existing VCs would be happy to put in all the money, but if they did, under the rules of the National Venture Capital Association, they couldn't write up the value of their Series A shares to the new, higher share price. However, if a new investor leads the round and agrees to the new, higher price, then they can show an unrealized increase in the value of their earlier investment, which will keep their limited partners (LPs) happy and help them raise their next fund.

So they find venture fund C, which agrees to invest 40% of the round, and venture capital funds A and B each invest 30% of the new round. Fund C insists on a new class of stock, Series B participating convertible preferred shares. The various preferences of the Series B shares take precedence over those of the Series A shares — the most recent money always takes priority over the previous investments. At $2/share, the company only has to sell 5 million shares to raise $10 million. Nearly all of the 1 million options in the original option pool have been granted to current employees, so the VCs authorize issuance of a further 1 million shares to the option pool so that the company can issue options to the next group of employees who'll be hired. Some of these option shares can also be issued to existing employees, especially those high-performers who are critical to the ongoing success of the company.

The Cap Table after the Series B round is shown in Table 4.

The VCs now own 71.8% of the company on an issued and outstanding basis and 59.6% on a fully diluted basis. The professor's share is down to 5.1% on an issued and outstanding basis and 4.3% on a fully diluted basis but the value of his shares has increased to $1 million. The pre-money valuation for the round was $13.5 million and the post-money value is $23.5 million.

INITIAL PUBLIC OFFERING

The early sales of the company's first product are going extremely well, so with revenues to report from its now validated first product, the company decides it is ready

Table 4: Cap Table after Series B Round

Price per share: $2.00

	Shares				Raised	%		Value
	Common		Series A	Series B		I&O	FD	
	Shares	Options						
Professor	500,000					4.7%	3.9%	$1,000,000
Postdoc A	250,000					2.3%	2.0%	$500,000
Postdoc B	250,000					2.3%	2.0%	$500,000
University	500,000					4.7%	3.9%	$1,000,000
CEO	1,000,000					9.3%	7.8%	$2,000,000
Seed investors	250,000					2.3%	2.0%	$500,000
Management Pool		2,000,000					15.7%	$4,000,000
VC Fund A			1,500,000	1,500,000	$3,000,000	27.9%	23.5%	$6,000,000
VC Fund B			1,500,000	1,500,000	$3,000,000	27.9%	23.5%	$6,000,000
VC Fund C				2,000,000	$4,000,000	18.6%	15.7%	$4,000,000
Total	2,750,000	2,000,000	3,000,000	5,000,000	$10,000,000	100%	100%	$25,500,000
Issued and outstanding	10,750,000							
Fully diluted	12,750,000							
Raised in this round	$10,000,000							
Cumulative raised	$13,210,000							
Pre-Money	$15,500,000							
Post-Money	$25,500,000							

to file for an initial public offering, or IPO[2]. It finds an investment banker who feels it can underwrite a sale of 8 million shares to the public at $8/share even though the company is not yet profitable (this is typical for biotechnology companies). Immediately before the public offering, all shares of Series A and B participating convertible preferred are converted into common shares, and the holders of the options all exercise their options so that they will be able to sell the shares and obtain long term capital gains tax treatment of their profit.

The Cap Table now looks very different, as shown in Table 5.

The public shareholders now own 40.5% of the company, the VC investors own 35.4%, the seed investors own 1.3% and the founders and management own 22.8%. There is only a single class of stock, common stock. Tje company is now back to where it started, with one class of stock with no preferences. Some companies that enter the public market attempt to maintain some preferred voting preferences but that is seldom done, and is beyond the scope of this article.

The $12 million invested by the 3 VC funds in the Series A and B rounds has increased to $56 million, with VC funds A and B showing a 5x return on the $4 million they each invested in the Series A and B rounds and VC fund C showing a 4x return on the $4 million it invested in the Series B round. The professor's ownership of the company is down to 2.5% of the company, but his shares are now worth $4 million. However, the money is not yet in the bank since hisas shares are not yet "liquid" as we cover in the next section.

LIFE AFTER THE IPO

The VCs, founders and management cannott sell their shares immediately. First, the underwriters will have imposed a "lock-up" of 6 months, during which none of the existing shareholders can sell their stock. The lock-up allows an orderly public market for the company's shares to develop. Second, the existing shareholders own unregistered shares — shares that have not been registered with the SEC. Only the public shareholders own registered stock at this stage and can sell it freely. Before the existing shareholders can sell their shares the shares need to

2 In reality, it is highly unlikely that the company will be able to go public after raising and investing so little. However, we will learn nothing new by going through Series C, D, E etc. rounds of VC financing, except that we would see founders and management getting diluted to the stage that the investors may start to give them options to get their shareholdings back up. VCs like to see the CEO not drop below 5% and the other "C" level members of the management team stay around 2%.

Table 5: Cap Table after IPO

Price per share: $8.00

	Shares	Raised	%		Value
	Common		I&O	FD	
	Shares				
Professor	500,000		2.4%	2.4%	$4,000,000
Postdoc A	250,000		1.2%	1.2%	$2,000,000
Postdoc B	250,000		1.2%	1.2%	$2,000,000
University	500,000		2.4%	2.4%	$4,000,000
CEO	1,000,000		4.8%	4.8%	$8,000,000
Seed investors	250,000		1.2%	1.2%	$2,000,000
Management Pool	2,000,000		9.6%	9.6%	$16,000,000
VC Fund A	3,000,000		14.5%	14.5%	$24,000,000
VC Fund B	3,000,000		14.5%	14.5%	$24,000,000
VC Fund C	2,000,000		9.6%	9.6%	$16,000,000
Public Investors	8,000,000	$64,000,000	38.6%	38.6%	$64,000,000
Total	20,750,000	$64,000,000	100%	100%	$166,000,000
Issued and outstanding	20,750,000				
Fully diluted	20,750,000				
Raised in this round	$64,000,000				
Cumulative raised	$77,210,000				
Pre-Money	$102,000,000				
Post-Money	$166,000,000				

be registered with the SEC. The VCs will have included the right for registration of their shares in their preferences, and hopefully management has negotiated "tag along" rights so that they can register some or all of their shares.

That said, a small amount of shares can be sold under Rule 144, the amount being related to the daily trading volume of the company's publicly traded shares.

ACQUISITION

An attractive alternative to an IPO is to consider selling the company to another, bigger company. The acquisition will either be paid for in cash or in shares of the acquiring company's stock, if the company is already publically traded. Acquisition is attractive because; (a) there is immediate liquidity since the purchase price is either paid in cash or through registered shares of the acquiring company; and. (b) an IPO is an expensive undertaking, and the underwriter commissions and legal and accounting fees will typically consume at least 10% of the funds raised.

However, from the management and founders' viewpoint there is a downside to an acquisition — remember the liquidation preferences associated with the preferred shares? If a company is acquired, the preferred investors will typically first receive their investment, and sometimes a multiple of their investment, out of the purchase price, and the balance will be distributed among all the shareholders, including the preferred shareholders, according to their shareholdings. In other words, the preferred shareholders get a "double dip".

So, our company accepts an acquisition offer at $7.20/share, 10% below the IPO share price, with both the Series A and the Series B round investors having agreed to a 1x liquidation preference as part of their original investments — i.e., they will get their original investment back off the top and then get their ownership percentages of the balance of the proceeds.

The Cap Table and how the proceeds stack up are shown in Table 6, together with how the various constituents' fare compared with the value created in the IPO (i.e., assuming that all the shares are ultimately sold at the IPO price).

The comparison column shows that the holders of common stock receive 77.2% of the amount they would have received in the IPO, while venture funds A and B receive 97.2% and venture fund C comes out ahead, receiving 102.2% of the IPO amount. The common stock holders are hit by the reduced sale price and also by the preferences. However, for venture funds A and B, the

Table 6: Cap Table after acquisition

Acquisition price: $91,800,000
Per share: $7.2

Liquid. Pref.	Price
Series A	1 x $1.00
Series B	1 x $2.00

	Shares				%		Proceeds					
	Common		Series A	Series B	I&O	FD	Preferences	Balance	Total	IPO	Δ	%
	Shares	Options										
Professor	500,000				4.7%	3.9%		$3,090,196	$3,090,196	$4,000,000	($909,804)	77.3%
Postdoc A	250,000				2.3%	2.0%		$1,545,098	$1,545,098	$2,000,000	($454,902)	77.3%
Postdoc B	250,000				2.3%	2.0%		$1,545,098	$1,545,098	$2,000,000	($454,902)	77.3%
University	500,000				4.7%	3.9%		$3,090,196	$3,090,196	$4,000,000	($909,804)	77.3%
CEO	1,000,000				9.3%	7.8%		$6,180,392	$6,180,392	$8,000,000	($1,819,608)	77.3%
Seed investors	250,000				2.3%	2.0%		$1,545,098	$1,545,098	$2,000,000	($454,902)	77.3%
Mgmt Pool		2,000,000				15.7%		$12,360,784	$12,360,784	$16,000,000	($3,639,216)	77.3%
VC Fund A			1,500,000	1,500,000	27.9%	23.5%	$4,500,000	$18,541,176	$23,041,176	$24,000,000	($958,824)	96.0%
VC Fund B			1,500,000	1,500,000	27.9%	23.5%	$4,500,000	$18,541,176	$23,041,176	$24,000,000	($958,824)	96.0%
VC Fund C				2,000,000	18.6%	15.7%	$4,000,000	$12,360,784	$16,360,784	$16,000,000	$360,784	102.3%
Total	2,750,000	2,000,000	3,000,000	5,000,000	100%	100%	$13,000,000	$78,800,000	$91,800,000	$102,000,000	($10,200,000)	90.0%
Issued & Outstanding	10,750,000											
Fully Diluted	12,750,000											

preferences almost compensate for the reduced per share price. With venture fund C, since they only invested in Series B at the higher per share price the preference that they receive on the Series B investment more than compensates for the reduced per share price.

In reality the net proceeds to the company from an IPO at $8/share and an acquisition at $7.20/share are likely to be pretty similar — underwriter commissions are likely to be 7-8% of the proceeds, and the legal costs of an IPO, particularly complying with Sarbanes-Oxley Act, and interacting with the SEC will be substantially higher than for an acquisition, so in reality the common shareholders would receive 85.5% of the IPO gains, the amount they lose to the preferences, while all of the VCs come out ahead; VCs A and B getting 108% and VC C receiving 113.6% of the IPO amount.

THE DARK SIDE — DOWN ROUNDS

The following scenario illustrates what happens when all does not go well for the company. Let us assume that they don't achieve the milestones set by their investors in their Series A financing, and as a result they are in serious danger of running out of money so their bargaining power is not very good. In these circumstances, they will not be able to bring a new investor on board, and the round will be held just with venture funds A and B. venture funds A and B are not happy since they can not mark up their investment and might actually have to mark it down — not a good thing forsomething their LPs will welcome. They still think the company is going to be successful, and are willing to put in more funds, but they extract their revenge. The company is still going to need $10 million to gear up to get to market, but it needs a further $1 million to cover the unexpected difficulties it has encountered in developing the lead product.

Venture funds A and B agree to invest the $11 million[3], but instead of agreeing to a $2/share price, they refuse to pay more than $0.60/share, plus they want a 3x liquidation preference! They will still agree to increase the option pool by 1 million shares. The company has no alternatives available to this offer, so it has to agree.

The Cap Table after the down round Series B is shown in Table 7.

The result is that the company has to issue over 18 million new shares and the professor's share has gone down to 2.2% on an issued and outstanding basis and 2.0% on a fully diluted basis, versus 5.1% and 4.3% in the base case scenario, and the value of his holdings has gone down from $1 million in the original scenario to $300,000. The investors now own 88% of the company on an issued and outstanding basis and 81% on a fully diluted basis. A down round is one reason why venture capitalists get called "vulture capitalists."

3 This is probably an unrealistic scenario — the company is much more likely to receive the $1 million in the next round to see if it can catch up, and then to get the $10 million in a subsequent round if it does. I assume it all comes in in one round to provide more of an "apples-to-apples" comparison and to magnify the impacts.

Table 7: Cap Table after down round Series B

Price per share: $0.60

	Shares				Raised	%		Value
	Common		Series A	Series B		I&O	FD	
	Shares	Options						
Professor	500,000					2.2%	2.0%	$300,000
Postdoc A	250,000					1.1%	1.0%	$150,000
Postdoc B	250,000					1.1%	1.0%	$150,000
University	500,000					2.2%	2.0%	$300,000
CEO	1,000,000					4.3%	4.0%	$600,000
Seed investors	250,000					1.1%	1.0%	$150,000
Management Pool		2,000,000					8.0%	$1,200,000
VC Fund A			1,000,000	9,166,667	$5,500,000	44.0%	40.5%	$6,100,000
VC Fund B			1,000,000	9,166,667	$5,500,000	44.0%	40.5%	$6,100,000
Total	2,750,000	2,000,000	2,000,000	18,333,333	$11,000,000	100%	100%	$15,050,000
Issued and outstanding	23,083,333							
Fully diluted	25,083,333							
Raised in this round	$11,000,000							
Cumulative raised	$14,210,000							
Pre-Money	$4,050,000							
Post-Money	$15,050,000							

IPO AFTER A DOWN ROUND — THE REVERSE SPLIT

Let us assume the company solves its problems with R&D, successfully develops its lead product, and starts sales. The investment bankers again feel they can take the company public and sell shares to individual investors. They want to price the shares at $8/share, but feel that the company has too many shares outstanding — over 25 million, versus less than 12 million in the original scenario. They therefore tell the company that they are going to have to do a 1:2 reverse split — i.e., shareholders surrender their share certificates to the company, and for every two old shares owned, they recieve one new share. The bankers tell the company that if they perform a reverse split, they will be able to sell 8 million shares to the public at $8/share, just as in the original scenario. The company wants to go public so that the investors and management can achieve liquidity, so they agree.

The Cap Table after the IPO with reverse split is shown in Table 8, together with a comparison with the outcome of the base case IPO.

The professor's shareholding is down to 1.2% vs. 2.5% in our base case scenario, and the value of his equity holding is now down to $2 million versus $4 million in the base case.

Figure 1 shows the build up in value of the company over time. Clearly the bulk of the value is created at the end of the process.

ACQUISITION AFTER A DOWN ROUND

The next scenario illustrates what happens if the company is acquired after a down round rather than going public. The key difference between this and the previous acquisition case that we considered is that as part of the punitive Series B financing, when the share price dropped

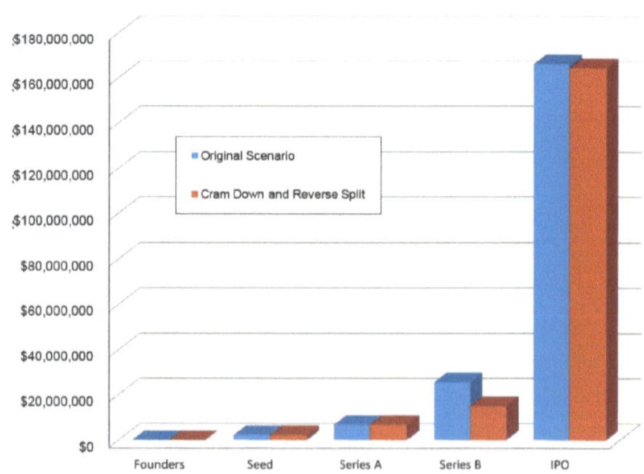

Figure 1: Build-up in company value over time

Table 8: Cap Table after IPO with reverse split

Price per share: $8.00
Reverse split: 1 for 2

						Original Scenario	
	Shares	Raised	%	Value		Value	Diff
	Common		I&O				
	Shares						
Professor	250,000		1.2%	$2,000,000		$4,000,000	($2,000,000)
Postdoc A	125,000		0.6%	$1,000,000		$2,000,000	($1,000,000)
Postdoc B	125,000		0.6%	$1,000,000		$2,000,000	($1,000,000)
University	250,000		1.2%	$2,000,000		$4,000,000	($2,000,000)
CEO	500,000		2.4%	$4,000,000		$8,000,000	($4,000,000)
Seed investors	125,000		0.6%	$1,000,000		$2,000,000	($1,000,000)
Management Pool	1,000,000		4.9%	$8,000,000		$16,000,000	($8,000,000)
VC Fund A	5,083,333		24.7%	$40,666,667		$24,000,000	$16,666,667
VC Fund B	5,083,333		24.7%	$40,666,667		$24,000,000	$16,666,667
VC Fund C						$16,000,000	($16,000,000)
Public Investors	8,000,000	$64,000,000	38.9%	$64,000,000		$64,000,000	$0
Total	20,541,667	$64,000,000	100%	$164,333,333		$166,000,000	($1,666,667)
Issued and outstanding	20,541,667					20,750,000	($208,333)
Fully diluted	20,541,667					20,750,000	($208,333)
Raised in this round	$64,000,000					$64,000,000	$0
Cumulative raised	$78,210,000					$77,210,000	$1,000,000
Pre-Money	$100,333,333					$102,000,000	($1,666,667)
Post-Money	$164,333,333					$166,000,000	($1,666,667)

by 40% to $0.60 per share rather than doubling to $2.00 per share, the investors also demanded and received a 3x liquidation preference. Since each venture fund invested $5.5 million in this round, they will each receive $16.5 million off the top of the acquisition proceeds, in addition to the 1x multiple of their $1 million investments in the Series A.

The Cap Table after the acquisition is shown in Table 9, together with a comparison with our acquisition base case.

The result is a massive shift of the proceeds from the common shareholders to the preferred. The founders receive less than a third of what they got in the base case, while the two venture funds receive double what they received in the base case (though, in fairness, they also each invested over 50% more — $6.5 million each versus $4 million.) This outcome again illustrates why venture capitalists are sometimes called "vulture capitalists."

ANTI-DILUTION

One of the emotive issues that always arises in negotiations with start-ups is that of anti-dilution. Everyone would like anti-dilution protection, but of course someone has to be diluted if new employees are to be hired or new investors brought into the company.

It is important to distinguish between two types of anti-dilution:

- The anti-dilution included in preference terms to protect early investors against down rounds; and
- The anti-dilution equity-ownership model frequently employed by universities.

Investor Protection Against Subsequent Down-Rounds

One of the preferences that will be in the preferred share investments will be anti-dilution protection. Anti-dilution protection comes in two flavors:

- Full Ratchet anti-dilution protection; and
- Weighted Average anti-dilution protection

Table 9: Cap Table after acquisition after down round

Acquisition price: $91,800,000
Per share: $7.2
Liquidation Preferences
Series A 1 x $1.00
Series B 3 x $0.60

	Shares				%		Proceeds					
	Common Shares	Options	Series A	Series B	I&O	FD	Preferences	Balance	Total	Base Case	D	%
Professor A	500,000				2.2%	2.0%		$1,132,226	$1,132,226	$3,090,196	($1,957,970)	36.6%
Postdoc B	250,000				1.1%	1.0%		$566,113	$566,113	$1,545,098	($978,985)	36.6%
Postdoc C	250,000				1.1%	1.0%		$566,113	$566,113	$1,545,098	($978,985)	36.6%
University	500,000				2.2%	2.0%		$1,132,226	$1,132,226	$3,090,196	($1,957,970)	36.6%
CEO	1,000,000				4.3%	4.0%		$2,264,452	$2,264,452	$6,180,392	($3,915,940)	36.6%
Seed investors	250,000				1.1%	1.0%		$566,113	$566,113	$1,545,098	($978,985)	36.6%
Management Pool		2,000,000				8.0%		$4,528,904	$4,528,904	$12,360,784	($7,831,881)	36.6%
VC Fund A			1,000,000	9,166,667	44.0%	40.5%	$17,500,000	$23,021,927	$40,521,927	$23,041,176	$17,480,750	175.9%
VC Fund B			1,000,000	9,166,667	44.0%	40.5%	$17,500,000	$23,021,927	$40,521,927	$23,041,176	$17,480,750	175.9%
VC Fund C										$16,360,784	($16,360,784)	
Total	2,750,000	2,000,000	2,000,000	18,333,333	100%	100%	$35,000,000	$56,800,000	$91,800,000	$91,800,000	$0	100.0%
Issued and outstanding	23,083,333											
Fully diluted	25,083,333											

The way these anti-dilution measures actually operate is that the conversion price of the preferred stock into common stock prior to an acquisition or IPO — which is normally set up as 1.0 to 1.0 is adjusted to a lower figure. So if the anti-dilution mechanism lowered the conversion rate of a round to say 0.8 to 1.0, the preferred shareholder would get 25% more common shares than they would otherwise have received.

Full ratchet anti-dilution protection

Full ratchet anti-dilution protection is draconian and it should be fairly easy to negotiate it away. In full ratchet anti-dilution protection, the price of earlier purchased shares is adjusted down to the latest price, and the number of shares is increased to the number that the earlier round investment would have purchased at this lower price. In our case, the Series B was priced at $0.60 per share, so the price of the Series A would be adjusted to convert at 0.60 shares per share of common stock. At a conversion ratio of 0.60 per share, Venture Fund A and B's original $1 million investments would each have been converted into 1,666,667 shares of stock, so an additional 666,667 shares would have been issued to both Venture Fund A and B. Table 10 shows the Cap Table after a Down Round Series B with full ratchet anti-dilution protection. The effect is to lower each common shareholders' ownership of the company by 10% (e.g., the professor goes from 2.2% to 2.0%), while venture funds A and B each increase 1%, from 44.0% to 44.4%.

Weighted average anti-dilution protection

Weighted average anti-dilution protection is less punitive to common shareholders, and it adjusts the price of earlier purchasers at a higher price by weighting the decrease in price by the amount of money raised at the higher and lower prices.

So, in our example, the conversion price of the Series A shares would be multiplied by:

number of shares actually issued / number of shares that would be issued at new lower price
or
(18,333,334+2,000,000) / (18,333,334+3,333,334)
or,
0.9385 shares per share of common stock

1,000,000 shares of preferred stock would convert into 1,065,557 shares at a conversion ratio of 0.941 to 1. Therefore 65,557 additional shares would be issued to each of venture funds A and B, a far cry from the 666,667 they would each receive under full ratchet anti-dilution.

Impact of anti-dilution protection

The effect of anti-dilution protection is to shift ownership from the common shareholders to the preferred. The case can be made that management deserves to be punished in this way, since they are responsible for the failure to achieve the agreed upon milestones, but the university may feel aggrieved to be punished in this way — after all, their technology is still the core of the company and they were not involved with the company's operations and hence failure to meet milestones.

Table 10: Cap Table after down round Series B with full ratchet anti-dilution protection

Price per share: $0.60

	Shares				Raised	%		Value
	Common		Series A	Series B		I&O	FD	
	Shares	Options						
Professor	500,000					2.0%	1.9%	$300,000
Postdoc A	250,000					1.0%	0.9%	$150,000
Postdoc B	250,000					1.0%	0.9%	$150,000
University	500,000					2.0%	1.9%	$300,000
CEO	1,000,000					4.1%	3.8%	$600,000
Seed investors	250,000					1.0%	0.9%	$150,000
Management Pool		2,000,000					7.6%	$1,200,000
VC Fund A			1,666,667	9,166,667	$5,500,000	44.4%	41.0%	$6,500,000
VC Fund B			1,666,667	9,166,667	$5,500,000	44.4%	41.0%	$6,500,000
Total	2,750,000	2,000,000	3,333,334	18,333,333	$11,000,000	100%	100%	$15,850,000
Issued and outstanding	24,416,667							
Fully diluted	26,416,667							
Raised in this round	$11,000,000							
Cumulative raised	$14,210,000							
Pre-Money	$4,850,000							
Post-Money	$15,850,000							

Universities have started including "anti-down round" protection clauses in their license agreements to address this issue.

University anti-dilution model

An alternative to the university being treated as a co-founder and receiving a significant equity stake — 20% in our base case — an approach universities frequently take is to say: "I don't care how much of the company I own now, I care how much I own after serious investors have valued the company by investing in it, so give me 5% and keep me at 5% until $5 million has been raised."

The advantages to the university are:

- It sounds less to the other founders and so is an easier sell; and
- The university doesn't have to worry about the company issuing additional Founders shares before investors come in and strictly limit the company's ability to issue additional shares.

Venture capitalists are familiar with these arrangements and as long as there is a clearly defined endpoint to the anti-dilution protection and the percentage ownership that is being protected is reasonable — e.g., 5% rather than 20% — such provisions will not be a barrier to the company raising funding.

Tables 11-13 show what the Cap Table would look like through Series A if the university negotiated to receive 10% with anti-dilution protection on a fully diluted basis to $3 million raised excluding the Seed Round.

The university would receive only 885 shares in the pre-split founders round, rather than the 2,000 shares in our base case.

These would become 221,250 shares following the 250 for one split prior to the Seed Round, plus a further 27,000 shares would need to be issued to bring the university back up to 10% after the Seed Round.

After the Series A, an additional 445,000 shares would need to be issued to bring the university back up to 10% on a fully diluted basis. At this point the anti-dilution protection is exhausted and the university will undergo the same dilution as other shareholders going forward.

The university therefore owns 693,250 shares, a 10% stake, after the Series A, versus 500,000 shares, a 7.4% stake, in our base case, and it is clear that 10% with anti-dilution protection to $3 million raised is worth considerably more than 20% of the founders round.

Tables 14-17 show what the Cap Table would look like through Series B if the university instead negotiated to receive 5% with anti-dilution protection on a fully diluted basis through $5 million raised excluding Seed Round.

The university would only receive 425 shares in the pre-split founders round.

These would become 106,250 shares after the split preceding the Seed Round, plus it would receive a further 11,000 shares to bring it back to 5% after the Seed Round.

The university would receive a further 210,000 shares to bring it back up to 5% after the Series A.

Table 11: Cap Table after founders round, 10% anti-dilution protection till $3 million raised

Price per share: $1.00

	Shares	Raised	%	Value
Professor	2,000	$2,000	22.5%	$2,000
Postdoc A	1,000	$1,000	11.3%	$1,000
Postdoc B	1,000	$1,000	11.3%	$1,000
University	885	$885	10.0%	$885
CEO	4,000	$4,000	45.0%	$4,000
Total	8,885	$8,885	100%	$8,885
Issued and outstanding	8,885			
Fully diluted	8,885			
Raised in this round	$8,885			
Cumulative raised	$8,885			

Table 12: Cap Table after Seed Round, 10% anti-dilution protection till $3 million raised

Price per share: $0.80
Split: 250 for 1

	Shares	Raised	%	Value
Professor	500,000		20%	$400,000
Postdoc A	250,000		10%	$200,000
Postdoc B	250,000		10%	$200,000
University	221,250		8.9%	$198,600
Anti-Dilution Shares	27,000	$108	1.1%	$221,600
CEO	1,000,000		40%	$800,000
Seed investors	250,000	$200,000	10%	$200,000
Total	2,498,250	$200,108	100%	$2,220,200
Issued and outstanding	2,498,250			
Fully diluted	2,498,250			
Raised in this round	$200,108			
Cumulative raised	$208,993			
Pre-Money	$2,020,092			
Post-Money	$2,220,200			

Table 13: Cap Table after Series A Round, 10% anti-dilution protection till $3 million raised

Price per share: $1.00

	Shares			Raised	%		Value
	Common		Series A		I&O	FD	
	Shares	Options					
Professor	500,000				8.4%	7.2%	$500,000
Postdoc A	250,000				4.2%	3.6%	$250,000
Postdoc B	250,000				4.2%	3.6%	$250,000
University	248,250				4.2%	3.6%	$693,250
Anti-Dilution Shares	445,000			$1,780	7.5%	6.4%	$695,000
CEO	1,000,000				16.8%	14.4%	$1,000,000
Seed investors	250,000				4.2%	3.6%	$250,000
Management Pool		1,000,000				14%	$1,000,000
VC Fund A			1,500,000	$1,500,000	25%	22%	$1,500,000
VC Fund B			1,500,000	$1,500,000	25%	22%	$1,500,000
Total	2,943,250	1,000,000	3,000,000	$3,001,780	100%	100%	$7,638,250
Issued and outstanding	5,943,250						
Fully diluted	6,943,250						
Raised in this round	$3,001,780						
Cumulative raised	$3,211,780						
Pre-Money	$4,636,470						
Post-Money	$7,638,250						

The $10 million raised in the Series B Round blows through the anti-dilution limit of $5 million, so to calculate how many shares the University should receive, we break the transaction down into two transactions — a $2 million investment to get to the $5 million anti-dilution limit and an $8 million investment to complete the round. Now, the option pool increases from 1,000,000 to 2,000,000 shares as part of the Series B Round, and the original agreement requires that the 5% be calculated on

Table 14 Cap Table after founders round, 5% anti-dilution protection till $5 million raised

Price per share: $1.00

	Shares	Raised	%	Value
Professor	2,000	$2,000	23.7%	$2,000
Postdoc A	1,000	$1,000	11.9%	$1,000
Postdoc B	1,000	$1,000	11.9%	$1,000
University	425	$425	5.0%	$425
CEO	4,000	$4,000	47.5%	$4,000
Total	8,425	$8,425	100%	$8,425
Issued and outstanding	8,425			
Fully diluted	8,425			
Raised in this round	$8,425			
Cumulative raised	$8,425			

Table 15 Cap Table after Seed Round, 5% anti-dilution protection till $5 million raised

Price per share: $0.80
Split: 250 for 1

	Shares	Raised	%	Value
Professor A	500,000		21.1%	$400,000
Postdoc B	250,000		10.6%	$200,000
Postdoc C	250,000		10.6%	$200,000
CEO	1,000,000		42.2%	$800,000
University	106,250		4.5%	$93,800
Anti-Dilution Shares	11,000	$44	0.5%	$208,800
Seed investors	250,000	$200,000	10.6%	$200,000
Total	2,367,250	$200,044	100%	$2,102,600
Issued and outstanding	2,367,250			
Fully diluted	2,367,250			
Raised in this round	$200,044			
Cumulative raised	$208,929			
Pre-Money	$1,902,556			
Post-Money	$2,102,600			

Table 16: Cap Table after Series A Round, 5% anti-dilution protection till $5 million raised

Price per share: $1.00

	Shares			Raised	%		Value
	Common		Series A		I&O	FD	
	Shares	Options					
Professor A	500,000				9%	8%	$500,000
Postdoc B	250,000				4%	4%	$250,000
Postdoc C	250,000				4%	4%	$250,000
CEO	1,000,000				18%	15%	$1,000,000
University	117,250				2.1%	1.8%	$327,250
Anti-Dilution Shares	210,000				3.8%	3.2%	$460,000
Seed investors	250,000				4%	4%	$250,000
Management Pool		1,000,000				15%	$1,000,000
VC Fund A			1,500,000	$1,500,000	27%	23%	$1,500,000
VC Fund B			1,500,000	$1,500,000	27%	23%	$1,500,000
Total	2,577,250	1,000,000	3,000,000	$3,000,000	100%	100%	$7,037,250
Issued and outstanding	5,577,250						
Fully diluted	6,577,250						
Raised in this round	$3,000,000						
Cumulative raised	$3,210,000						
Pre-Money	$4,037,250						
Post-Money	$7,037,250						

a fully diluted basis, so does the university get its 5% of the extra 1,000,000 shares in the option pool or not? This is a business issue, not a legal matter, and the university should specify in the term sheet that an increase in the option pool is considered to occur before the preferred shares are issued to remove any ambiguity on this issue.

In our case, I have assumed that the university did include this issue in the term sheet. Table 17 shows the Cap Table after the complete Series B. The university receives an extra 100,000 shares to get it to 5% on a fully diluted basis after the option pool is increased and $2 million of Series B Preferred is issued. The remaining $8 million investment takes the university's ownership down to 3.4% on a fully diluted basis at the end of the round.

Table 18 shows how the three approaches compare. Although the initial ownership percentages sound very different — 20%, 10% and 5% — the end results are not that different. 10% protected to $2 million results in the university ultimately owning almost 40% more shares than in the case of an unprotected 20%, while 5% protected to $5 million results in an ownership that is only 15% less than an unprotected 20%.

SUMMARY

This article has shown how the relative ownership shares of a start-up company evolve over time. It has also shown that the ultimate ownership by the various parties is going to be determined by the success of the company, and by careful management of the fund raising strategy — achieving value added milestones prior to major rounds of financing will preserve value for common shareholders.

The article also shows the value of non-dilutive funding — grants or partnerships where another party contributes services in kind. Suppose the initial $3 million product development phase had been funded by a grant from the federal government or a foundation — rather than through the Series A Round — then the common shareholders would not have suffered that particular 50% dilution and if the subsequent rounds of financing had remained the same, the founders' ultimate ownership share would have been double what it actually was.

Many company founders instead spend an inordinate amount of time worrying about dilution. Their energy would be more effectively used in focusing on value creation and the amount of money that can be made if the company is successful rather than trying to negotiate anti-dilution protection for themselves.

Table 17: Cap Table after Series B Round, 5% anti-dilution protection till $5 million raised

Price per share: $2.00

	Shares				Raised	%		Value
	Common		Series A	Series B		I&O	FD	
	Shares	Options						
Professor A	500,000					4.7%	3.9%	$1,000,000
Postdoc B	250,000					2.3%	2.0%	$500,000
Postdoc C	250,000					2.3%	2.0%	$500,000
CEO	1,000,000					9.4%	7.9%	$2,000,000
University	327,250					3.1%	2.6%	$854,500
Anti-Dilution Shares	100,000					0.9%	0.8%	$700,000
Seed investors	250,000					2.3%	2.0%	$500,000
Management Pool		2,000,000					16%	$4,000,000
VC Fund A			1,500,000	1,500,000	$3,000,000	14%	12%	$3,000,000
VC Fund B			1,500,000	1,500,000	$3,000,000	14%	12%	$3,000,000
VC Fund C				2,000,000	$4,000,000	19%	16%	$4,000,000
Total	2,677,250	2,000,000	3,000,000	5,000,000	$10,000,000	72%	76%	$20,054,500
Issued and outstanding	10,677,250							
Fully diluted	12,677,250							
Raised in this round	$10,000,000							
Cumulative raised	$10,210,000							
Pre-Money	$10,054,500							
Post-Money	$20,054,500							

Table 18: Comparison of university shareholdings under 3 negotiating models

Shares held by Univ after	Negotiating Model		
		Anti-Dilution	
	20%	10%/$3 mm	5%/$5mm
Founders	2,000	885	425
Seed	500,000	248,250	117,250
Series A	500,000	693,250	327,250
Series B	500,000	693,250	427,250

ss of Biotechnology

www.ingramcontent.com/pod-product-compliance
Ingram Content Group UK Ltd.
Pitfield, Milton Keynes, MK11 3LW, UK
UKHW060011051026
15591UKWH00002B/155
* 9 7 8 1 9 3 4 8 9 9 3 5 9 *